MAD
SCIENCE

Experiments You Can Do At Home, But STILL Probably Shouldn't 2

THEODORE GRAY

BLACK DOG
& LEVENTHAL
PUBLISHERS
NEW YORK

PUBLISHED BY
Black Dog & Leventhal Publishers, Inc.
151 West 19th Street
New York, NY 10011

DISTRIBUTED BY
Workman Publishing Company
225 Varick Street
New York, NY 10014

Manufactured in China

Cover and interior design by **Matthew Riley Cokeley**

Cover photograph: **Mike Walker**

ISBN-13: 978-1-57912-932-3

Library of Congress Cataloging-in-Publication Data available on file.

Contents

Introduction

When *Popular Science* asked me to write a monthly column for them back in 2002, my first reaction was to assume that they were confused. Asking me to write a single article, maybe, but a regular column when I'd never done anything like that before? That would be nutty.

Around the time I'd finished the second article for them, I did the math for how long I'd need to keep at it before I had enough columns to put together into a book. I figured fifty columns would be about right, and that meant about four or five years. So I started telling people that I planned to come out with a new book every five years like clockwork. Being sensible has never been one of my strong points.

But here we are, ten years later. My son is within half an inch of my height, my oldest daughter just turned sixteen, my little baby intends to move to France just as soon as legally possible, and my second book of collected *Popular Science* columns is finished. Life is a ride, and it's a privilege to be on it long enough to see a few crazy notions play out just like you had no right to expect.

I have no idea if there will be a third, fourth, or fifth *Mad Science* book, but so far I'm batting two for two. In the mean time, I think you're going to like this one. Capturing the photograph of a bullet going off in cutaway view is, in my opinion, one of the best things I've ever done, and the turkey fireball is almost as amazing as the smell it gave off. And, of course, the bacon lance has become something of a classic demonstration: I even got on a Penn & Teller show with that one!

If you read the first *Mad Science* book you know what you're in for, and rest assured this one includes all new demonstrations: There's no overlap whatsoever between the two books, except for the safety section that follows here. (Creating the photographs for that section was dangerous enough that I really didn't want to do another one like it. Getting myself injured while filming a safety section would just be too ironic.)

Please enjoy responsibly, have fun, and I'll see you in another five years (or next month in the magazine)!

Theodore Gray
December, 2012

Real warnings vs. the-lawyer-made-us-do-it warnings

It makes me cringe when I see warnings to wear gloves and safety glasses while working with baking soda. It's called crying wolf, and it's deeply irresponsible, because it makes it that much harder to get through to people about real dangers.

So I'm not going to do that. If you promise to listen, I promise to tell you the truth about where the real dangers are.

Some of the experiments in this book I would have let my kids do unsupervised when they were 10 years old (if not for the monumental messes that would lead to). If you're pouring a cold sodium acetate solution into a bowl, you are not going to get hurt, at least not by the sodium acetate. It's actually less toxic than common table salt, so unless you keep the salt in your house locked up and wear safety glasses for breakfast, you don't need to worry about sodium acetate.

Some other chemicals, however, are not your friends. Chlorine gas kills, and you hurt the whole time you're dying. Mix phosphorus and chlorates wrong and they blow up while you're mixing them. (I have a friend who still has tiny slivers of glass coming out of his hands twenty years after he made that particular mistake.)

Every chemical, every procedure, every experiment has its own unique set of dangers, and over the years people have learned (the hard way) how to deal with them. In many cases the only way to do an experiment safely is to find a more experienced person to help. This is not book-learning, it's your life at stake and you want someone by your side who knows what they are doing. There is an unbroken chain of these people leading right back to the first guy who survived, and you want to be part of that chain.

When I do an experiment that looks crazy I either have someone with me who's done it before, or it's something that I've worked my way up to slowly and carefully. I build in layers of safety, and I make sure that if all else fails I have a clear path to run like hell (and of course I wear glasses *at all times*).

I have never been seriously hurt by a chemical, and luck is not a factor in that. Don't make it a factor in your own safety either.

Should you actually try these experiments?

"Don't try this at home, kids!" Depending on your personality, that's either a warning or an invitation. I hate it because it tells people to be helpless—to believe that they are not smart enough, competent enough or persistent enough to do what "the experts" can do.

At the same time, it frightens me to think of someone picking up this book and ending up dead, burned or blind because of something I wrote, or a warning I didn't write. Some of these experiments would be just plain nuts for you to try. Seriously nuts.

Why nuts for you and not for me? Because each of us has a particular set of talents, experiences, friends and equipment. I do only things I know I can do safely. The things I didn't think I could do safely are not in this book, because I didn't do them.

For example, I saw a video of some guys who have learned to jump off huge cliffs wearing tiny wingsuits. They soar down the side of the mountain inches away from the ground and pull their parachutes at the last possible second. Are they nuts? Actually not; the ones who have survived this sport (many have not) are cautious people in their own slightly insane way. They started out trying to stay as far away from the cliff face as possible, until that got "boring."

A couple of the experiments in this book are in that category: things you can do safely only by edging up to them slowly and learning from the mistakes of others. They are not beginner experiments, just as jumping off a cliff in a wingsuit is not beginner skydiving.

Which brings me to an important point:

THIS BOOK DOES NOT TELL YOU ENOUGH TO DO ALL OF THE EXPERIMENTS SAFELY!

Some of the experiments you should be able to do safely using just the instructions in this book, combined with common sense and a modest amount of effort. But in many cases the steps are not detailed enough to allow you to do the experiment. They are there simply to illustrate in a general way how the experiment is done. A lot of experience is needed to fill in the blanks.

Please be honest with yourself in assessing whether you have the knowledge and experience needed before trying any of the experiments for real. Your safety depends on it, just as my safety depends on my knowing that, as fun as it might look, I should not jump off a cliff in a wingsuit anytime soon.

If you never read any warnings, please read this:

WEAR SAFETY GLASSES!

Nearly every experiment in this book has the potential to blind you. You have only two eyes, and they're close to each other: One splash of acid, and you're shopping for a cane.

I'm lucky to be so nearsighted that I have no choice but to wear glasses all the time. If you aren't, then you need to make the effort to get a good, comfortable pair of safety glasses. Not the cheap, crappy kind you're not going to wear, but some good ones that won't scratch and fog up all the time. They're only about $10 at a good home center or hardware store. Buy several so you can always find a pair. Wear them. Please, for my sake, wear them because I really, really don't want to get a letter from the mother of a kid who will never see his mother again.

Acknowledgments

As much as I would like to take credit as the genius cranking out crazy experiments single-handedly, this book is, of course, the result of many people's efforts. First and foremost I must acknowledge the huge contributions of my editors at *Popular Science*, including Mark Jannot, Mike Haney, Doug Cantor, Trevor Thieme, and Dave Mosher, and the book's designer, Matthew Cokeley. Without their diligent efforts, both during the original writing and editing of the columns and the assembly of this book, I would be just another disgruntled mad scientist posting on my blog.

The whole endeavor was set in motion by Mark Jannot, who first emailed me out of the blue asking if I'd like to write a monthly column for *Popular Science*, based on nothing more than my website *periodictabletable.com*. Like I'm going to say no? Mark set the tone for the column and patiently trained me to write only 400 words about 4,000-word topics.

The outstanding photographers who have worked with me over the years must receive a lot of credit for this book, which is as much about beautiful photography as anything. Mike Walker shot more columns than any other, and is actually starting to get used to the idea of photographing something flammable and potentially explosive each month. Jeff Sciortino, Rory Earnshaw and Chuck Shotwell were great to work with, hopefully their memories of the job are not too nightmarish. And while I have not had the privilege of working with the *PopSci* staff photographer John Carnett, his advice and support has been most helpful. My sidekick

Nick Mann (originally hired as a lackey, but he got promoted) shot video for many of the experiments.

For outstanding lunches during the long, grueling hours spent photographing the demonstrations in this book, I thank Koatie D. Pasley, who, despite the excellence of her cooking, should probably not open a business called "Koatie's Klassic Katering", because that would just not be a good idea on so many levels.

A number of scientists have provided invaluable, possibly lifesaving advice, including Triggvi Emilsson, who gave me the idea for my very first column on making ice cream with liquid nitrogen, and his colleagues Tim Brumleve and Sherwin Gooch. Valuable advice has also come from Ethan Currens, Blake Ferris, Simon Field, Bert Hickman, Gert Meyers, Jason Stainer, Bassam Z. Shakhashiri, Hal Sasabowski and Nick Younes. Column ideas were contributed by all of the above, as well as Chris Carlson, Niels Carlson and Oliver Sacks.

I thank Marcus Wynne for finding Beverly Martin at *agentresearch.com* who found me my agent James Fitzgerald, who I think for finding me my publisher, Black Dog & Leventhal, and my editor Becky Koh, who I thank for believing in the book even after she realized that it came with a crowd of people who all think they know how to do her job for her.

And, finally, I thank Jane Billman, Nina Paley and my children, Connor Gray, Addie Gray, and Emma Gray (listed in order of height) for putting up with me and helping with some of the experiments (but only, I would like to assure the authorities, when they were perfectly safe). And in case any of you are paying attention and compare these acknowledgments to the one from the first edition of *Mad Science*, yes, the order of children has changed. Connor grew. A lot.

Let There Be Light

A Light Mystery

Although we've long seen LEDs glow, we haven't always known why it happens

THE FIRST LIGHT-EMITTING DIODES went on sale in 1962, and you could have any kind you wanted as long as it was dim and red. Green, yellow and orange came next, but blue LEDs didn't debut until 1989. So it may surprise you that the first LEDs, discovered in 1907, included blue—and were made of sandpaper.

Well, not exactly sandpaper, but with the same material a lot of sandpaper uses, which is synthetic silicon carbide (carborundum). If you touch two needles to the surface of a crystal of silicon carbide and run electricity through them, you will sometimes see a very faint-colored glow. Silicon carbide is a semiconductor, and the needles on the surface create a diode, a device that allows electricity to flow in only one direction, so it really is a light-emitting diode.

When radio-development pioneer Henry Joseph Round noticed this glow in 1907, he published a short paper asking if anyone else had seen this and could explain it. No one had a clue.

The first commercially practical LEDs didn't arrive until a quantum-mechanical model for semiconductors allowed engineer Nick Holonyak, Jr., to design one with just the right electrical properties to create usable light.

Science is full of things you can see with your own eyes yet for which, even today, there is no satisfactory explanation. For instance, a compass needle always points north. You might know this happens because the Earth's magnetic field is oriented roughly along its axis of spin. But why does the Earth have a magnetic field, and why does it point north? No one knows. We can see it, describe it, and measure it, but we can't explain it.

BLUE LIGHT SPECIAL A nine-volt battery creates a tiny patch of blue light in a silicon carbide crystal. This is not a spark; it's the same electroluminescent effect that drives all LEDs.

> "Science is full of things you can see with your own eyes yet for which, even today, there is no satisfactory explanation."

Creating a visible amount of light using a silicon carbide crystal is not terribly difficult, but it is quite finicky. I had to keep trying out different points using two sharp needles, and didn't get anything more than a very small, barely visible glow. I tried several different silicon carbide crystals that I found on eBay, where they are readily available. None seemed particularly better than any other. To hold the needles, I used one of those positioning aids (available at Radio Shack) that has two alligator clips mounted on an adjustable base. I figured out after the battery started getting hot that I needed to insulate the needles from the alligator clips using electrical tape, otherwise there is a short circuit through the positioning device!

CRYSTAL GLOW Large silicon carbide crystals like this (sometimes available for sale on eBay) grow inside blast furnaces from a combination of the silica lining and the carbon in the coal that fires the furnace.

An Old Flame

Calcium carbide reactions can light up a room—or fill it with noise

PUSHING BACK THE NIGHT with light of our own making was the first and greatest of humankind's achievements. What a thrill it must have been to discover that the setting sun no longer had to mean darkness and fear. We've come a long way since that first campfire, but it's just recently that technology has topped the most advanced form of open-flame light. As a portable light source, only the LED flashlight is superior to the carbide miner's lamp, which had been the standard for brightness, weight and reliability since the early 1900s.

Carbide lamps have an upper chamber full of water and a lower chamber full of rocklike calcium carbide (CaC2). A valve lets the water drip onto the CaC2 at a steady rate. When water hits carbide, it produces acetylene gas, which is directed to a nozzle in a parabolic reflector and then lit manually. Burning acetylene is extremely bright, and a cave explorer can keep a light going for days with just some water and a bag of what look like ordinary stones.

Mixing acetylene with air also creates very loud bangs. This is what made a certain toy cannon so attractive when I was a kid. Powered by "Bangsite" (the toy maker's name for powdered calcium carbide), the cannons had a breach-load mechanism that dumped the compound into a water reservoir as it was locked in place. A firing plunger lit the gas with a flint.

I never did get one as a kid, and when I showed my dad the one I got for this column, he said that he'd always wanted one too! So kids, take heed: When pestering your parents for loud, dangerous toys, sell it as a chance to finally realize their childhood desires.

"The carbide miner's lamp... had been the standard for brightness, weight and reliability since the early 1900s."

Acetylene flame

Water chamber

JUSTRITE
MADE IN U.S.A.
PAT. MAY 7. 1912
PAT. OCT. 28. 1913
PAT. NOV. 23. 1915
OTHERS PENDING

Carbide chamber

LONG-LASTING LIGHT More like a blow torch than a candle, the flame from a carbide miner's lamp is almost impossible to blow out.

How I Did It

Calcium carbide "big bang" cannons are readily available on eBay, along with the granular carbide you need to operate them. They work great, but do wear hearing protection, and whatever you do, never plug up the barrel or try to have it shoot a projectile. The problem isn't that the projectile might do damage, it's that the barrels of these things are nowhere near strong enough to contain the force of the explosion if they aren't open on one end. If the barrel explodes, you are going to have a bad day.

Solid lumps of calcium carbide are also available on eBay, and can be used to create flammable acetylene just by putting them in a glass of water or on a block of ice, as I did. The bubbles will keep themselves lit as long as you have enough calcium carbide to create a steady stream of bubbles. Collecting the gas is not a good idea: Acetylene is too explosive to mess with this way.

Finally, you can also easily get antique carbine miner's lamps on eBay, but I would caution against this. I got several, and all of them had problems with the seals around the canister where water and carbide are mixed. These problems led to jets of flame shooting out of parts of the lamp where no jets of flame are meant to. The potential for disaster is somewhat limited by the fact that flammable acetylene isn't stored in the canister, it's being generated a bit at a time. But it's still pretty alarming, and one would want to carefully restore the seals before using one of these lamps for any length of time.

NOISEMAKERS A Big Bang toy cannon mixes acetylene gas with air when a stirrer, providing a tremendous bang.

REAL DANGER ALERT

Faulty seals on old carbide lamps or Big Bang toy cannons bought online can allow acetylene to escape from the chamber, resulting in fire where it does not belong. Also, cannons can cause hearing loss if misused.

Practical Knowledge

Fatal Attraction

Magnets don't have to be big to produce deadly force

IN THE PAST, MAGNETS WERE nothing to fear. The small ceramic type long used on refrigerators were barely strong enough to hold up a piece of paper. The same size magnet today can kill you.

Every electron in a material has a spin that creates a tiny magnetic field around it. Normally these electrons spin in random directions, canceling each other out, but in permanent magnets, some of the electrons are locked into alignment, producing an overall magnetic field. The stronger this lock-in, the stronger the magnet.

Powerful neodymium-iron-boron magnets are used in everything from jewelry to motors. With the development of new alloys and processing methods, they are getting stronger and stronger, to the point that even very small ones can be dangerous. If you swallowed two separately and they were to find each other, they could puncture your intestinal walls and cause a fatal infection.

Larger magnets, like the two-by-two-by-one-inch monsters I used in this demonstration, require careful planning just to carry from one room to another. Let one get too close to a steel door frame, and it can crush your hand. The two I used hug each other with 520 pounds of force.

"Two neodymium-iron-boron magnets hug each other with 520 pounds of force."

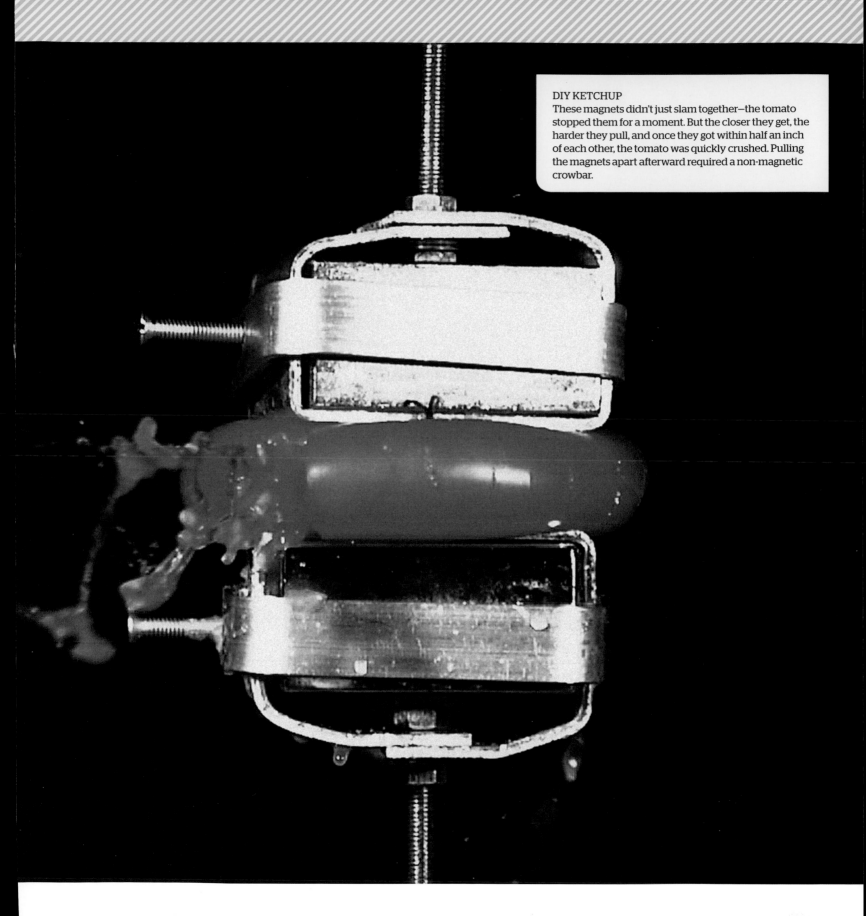

DIY KETCHUP
These magnets didn't just slam together—the tomato stopped them for a moment. But the closer they get, the harder they pull, and once they got within half an inch of each other, the tomato was quickly crushed. Pulling the magnets apart afterward required a non-magnetic crowbar.

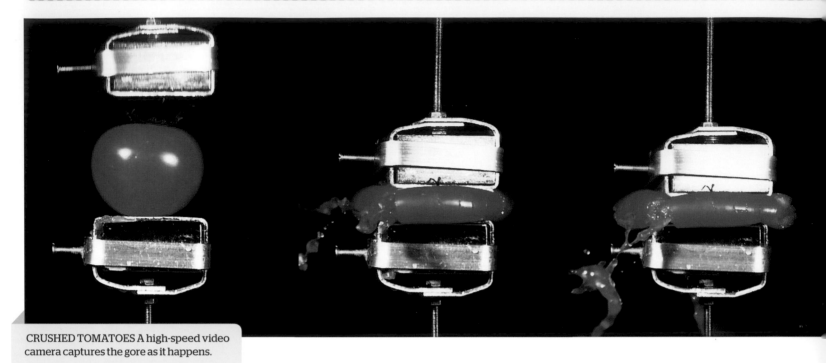

CRUSHED TOMATOES A high-speed video camera captures the gore as it happens.

How I Did It

Neodymium-iron-boron magnets do, however, have one weak spot: Heat them above 175°F, and the electron spins are knocked out of alignment, permanently destroying the magnetism. So if you ever find yourself with two of them clamped down on your finger, you could hold your hand in boiling water for a few minutes. But try to find a pry bar first.

It is somewhat alarming just how easy it is to buy huge super-magnets. My advice is, don't do it. Even small ones can hurt you and ones the size I used are seriously, seriously dangerous. I am actually considering heating mine up to destroy the magnetism just because they are too scary to keep around.

To do this demonstration I needed to let two of these beasts smash into each other, and then pull them back apart again: Not easy when the combined pulling force was over 500 pounds. I built a very sturdy rectangular frame (kind of like a picture frame) out of 2"x3" (non-magnetic) thick-wall aluminum square tubing. Then I built an aluminum cage for each magnet and attached a (non-magnetic) stainless steel threaded rod to each cage. One rod was permanently screwed to one side of the frame, and the other rod stuck through a slightly over-sized hole in the other side of the frame, allowing the second caged magnet to slide up and down. A nut and washer on the outside end of the second rod allowed me to use a (non-magnetic) bronze crowbar to lever the magnets apart after they hit each other by pulling on the rod. (By the way, if you ever find yourself, as I did, needing to buy a

 REAL DANGER ALERT

High-strength NdFeB magnets aren't toys. Even small ones can crush people's fingers. Do not actually try holding your hand in boiling water to remove the magnets.

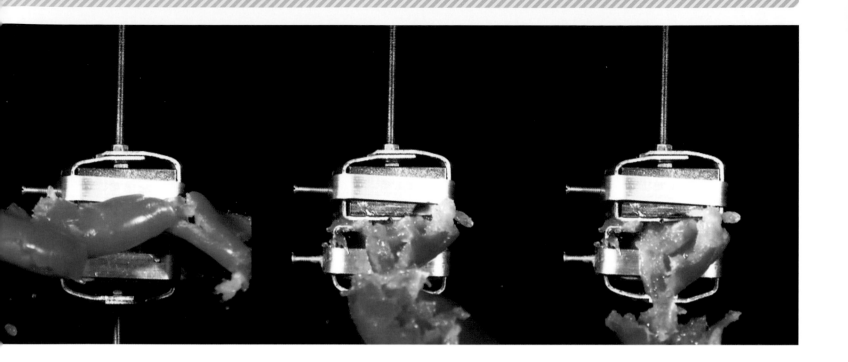

bronze crowbar in London, good luck. This kind of thing turns out to be much easier to get in America, land of the well-stocked hardware store.)

To be fair, the aluminum cages kept the magnets separated enough so that the pulling force required to separate them was probably closer to 200 pounds than 500, but it was still an impressive struggle to get them reset for the next shot. While I was working on the apparatus I always kept the magnets surrounded by blocks of Styrofoam a few inches thick. Only Styrofoam, you may ask, is that really strong enough? Yes, because the magnetic force only becomes strong once the magnets are quite close to each other. At six inches apart they pull enough to hold themselves up, but not enough to require anything strong to keep them separated.

We shot the footage using a 1200 frame-per-second high speed camera, which was definitely required given how fast and how forcefully the magnets slam together. There was tomato splattered everywhere that day. And Kiwi, and banana, and grapes, and just about any other fruits we could find for sale by the street vendors around East Acton. Tomatoes were the fruit of choice in the end due to the satisfying resemblance to blood and guts. (Yes, tomatoes are a fruit, look it up if you don't believe me.)

Can't Take the Heat

A change to Pyrex glass had far-reaching and unforeseen effects

MOST PEOPLE PROBABLY don't think of Corning as a crime-fighting company, but when it sold its Pyrex brand to World Kitchen in 1998, the company accidentally made the illegal manufacture of crack cocaine more difficult—a fascinating example of unintended consequences.

Ordinary glass shatters if it's heated too quickly: Pour boiling water into a common flint-glass tumbler, and it's likely to fall apart seconds later. The glass on the inside expands when it gets hot, putting stress on the cold glass on the outside. When the stress gets too great, it cracks.

Pyrex, which originally was always borosilicate glass, solved this problem by adding boron to the silica (quartz), the main ingredient in all glass. Boron changes the atomic structure of glass so it stays roughly the same size regardless of its temperature. Little thermal expansion means little stress. Thus borosilicate glass withstands heat not because it's stronger, but because it doesn't need to be stronger.

When World Kitchen took over the Pyrex brand, it started making more products out of pre-stressed soda-lime glass instead of borosilicate. With pre-stressed, or tempered, glass, the surface is under compression from forces inside the glass. It is stronger than borosilicate glass, but when it's heated, it still expands as much as ordinary glass does. It doesn't shatter immediately, because the expansion first acts only to release some of the built-in stress. But only up to a point.

One unfortunate use of Pyrex is cooking crack cocaine, which involves a container of water undergoing a rapid temperature change when the drug is converted from powder form. That process creates more stress than soda-lime glass can withstand, so an entire underground industry was forced to switch from measuring cups purchased at Walmart to test tubes and beakers stolen from labs. Which just goes to show, if you think you know all the consequences of your decisions today, you're probably wrong.

BREAKING STORY Because they're now made of soda-lime glass, Pyrex measuring cups can shatter violently when exposed to a sudden and extreme temperature change.

> **"Boron changes the atomic structure of glass so it stays roughly the same size regardless of its temperature."**

How I Did It

Not all measuring cups will work because some of them are in fact genuine borosilicate glass, which will not shatter under these conditions. Anything with the "pyrex" brand in all lower case should break properly. For those, the measuring cup needs to be heated up for a good 20–30 seconds with a plumbing torch before it will reliably shatter when a drop of water is dribbled into it.

We used a sound trigger to capture the shot a fraction of a second after the cup shattered: It happens too fast for human reaction times. There turns out to be quite a lot of variability in how measuring cups break: Sometimes they break into large pieces, other times they break into lots of little pieces. This indicates very poor quality control of the heat tempering process that is meant to be a substitute for using real borosilicate glass. If they were properly tempered, every measuring cup would burst into a million small pieces.

REAL DANGER ALERT

Do not try this demonstration yourself. All glass can shatter when exposed to drastic temperature changes, including consumer-grade Pyrex glassware, which has a fairly low resistance to thermal shock.

Gone in a Flash

Diamonds are for, well, a couple seconds

I CAN'T STAND DIAMONDS. No, really, they just tick me off, because nearly everything about them is a lie. Diamonds are neither rare nor intrinsically valuable nor uniquely romantic. Those are ideas invented by the diamond industry. And no, despite what the ads tell you, diamonds are not forever. They are flammable and will burn brightly with a little help from a torch. This makes perfect sense when you consider that they are made of pure carbon, which reacts with oxygen to form carbon dioxide ("reacts with oxygen" just being another way of saying "burns").

Diamond has one legitimate claim to fame: It is still, as far as we know, the hardest substance. Despite its hardness, though, the chemical bonds that hold the carbon atoms in diamond together are actually weaker than those that hold together the other common form of pure carbon, graphite. The difference is that in diamond the bonds form an inflexible, three-dimensional lattice, whereas in graphite the atoms are tightly bonded into sheets. But those sheets can slide easily against each other, making graphite soft and slippery.

It is bond strength, not hardness, that determines how easily oxygen can attack and burn a material, allowing me to burn a diamond in a pool of liquid oxygen resting in a block of graphite.

If your house burns down with the family jewels inside, you can collect the pools of melted gold, but the diamonds will be gone in a puff of CO_2. Cheaper, more attractive stones, such as cubic zirconia and synthetic ruby and sapphire, are made of refractory metal oxides that easily withstand the same heat. So it's actually mall trinkets, not diamonds, that are forever.

"Diamonds are neither rare nor intrinsically valuable nor uniquely romantic. Those are ideas invented by the diamond industry."

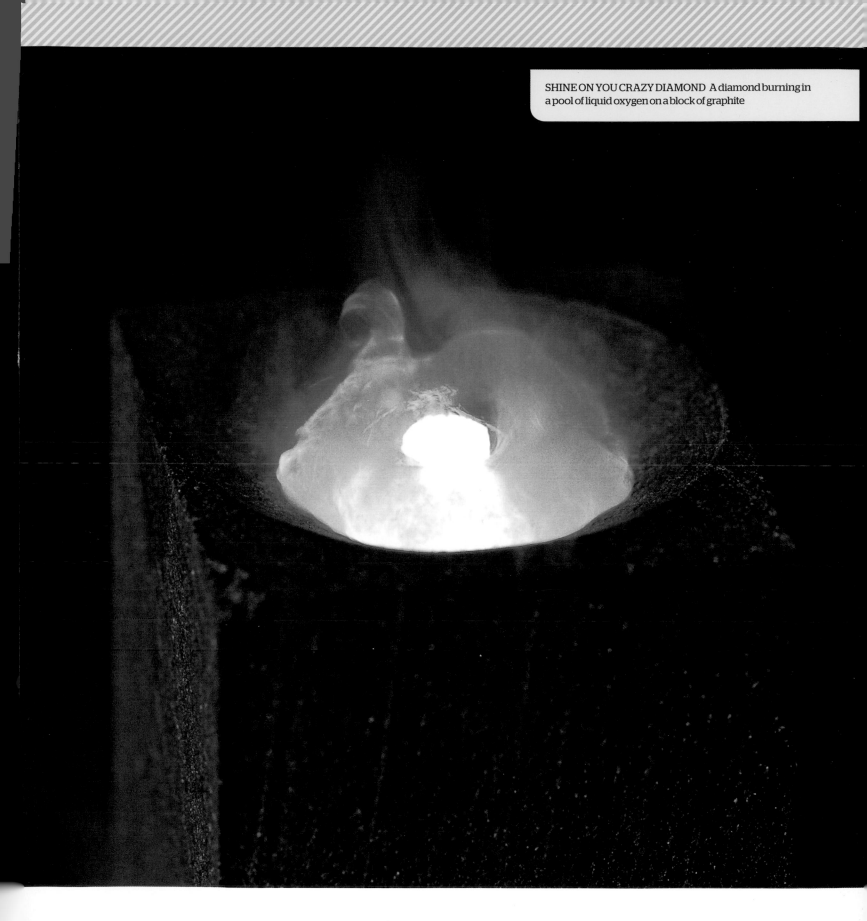

SHINE ON YOU CRAZY DIAMOND A diamond burning in a pool of liquid oxygen on a block of graphite

How I Did It

Diamonds can be made to burn in several different ways. I used liquid oxygen because in the end that gave the most pleasing photographic result, but oxygen gas from a welding cylinder works well too. The diamond is lit with a hot torch (hydrogen or acetylene will work) then a gentle stream of pure oxygen is blown at the diamond to keep it burning. You have to be careful not to blow too hard or the diamond will fly away and set something nearby on fire.

It took a while to figure out what to put the liquid oxygen and diamond in to get a good photograph. I tried a test tube, but it kept fogging up due to the coldness of the liquid oxygen. In the end I made a miniature graphite crucible by drilling a shallow hole in the end of a square stick of solid graphite (available on eBay). You might think that the oxygen would immediately evaporate on contact with a block of warm graphite, but actually the Leidenfrost effect (see p. 84) keeps it floating on an insulating cushion of gas.

Kids, be sure to ask your parents first before you burn up grandma's wedding ring.

CHEAP ROCKS The bean counters at PopSci let me burn only low-grade "Congo cube" and semitransparent diamonds, available on eBay for $50 to $300. The nice looking diamond ring is actually an $8 cubic zirconia ring from Walmart.

REAL DANGER ALERT

Don't try this demonstration at home. Diamonds can burst violently when heated. (We ruined an expensive camera lens with diamond bullets that could easily have taken an eye out.)

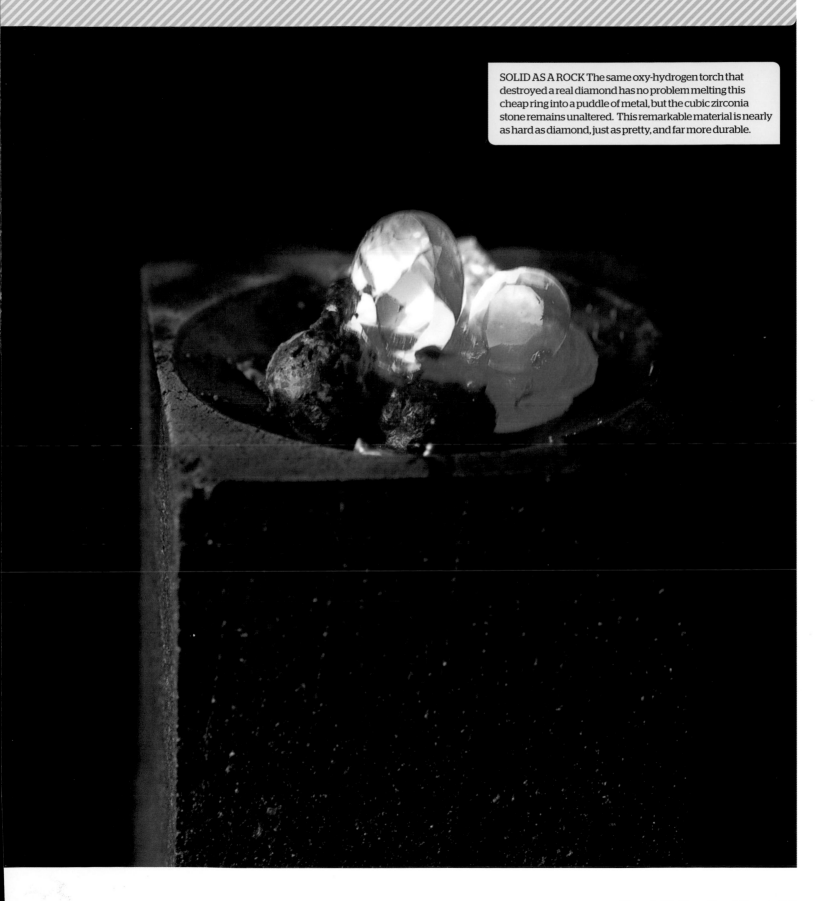

SOLID AS A ROCK The same oxy-hydrogen torch that destroyed a real diamond has no problem melting this cheap ring into a puddle of metal, but the cubic zirconia stone remains unaltered. This remarkable material is nearly as hard as diamond, just as pretty, and far more durable.

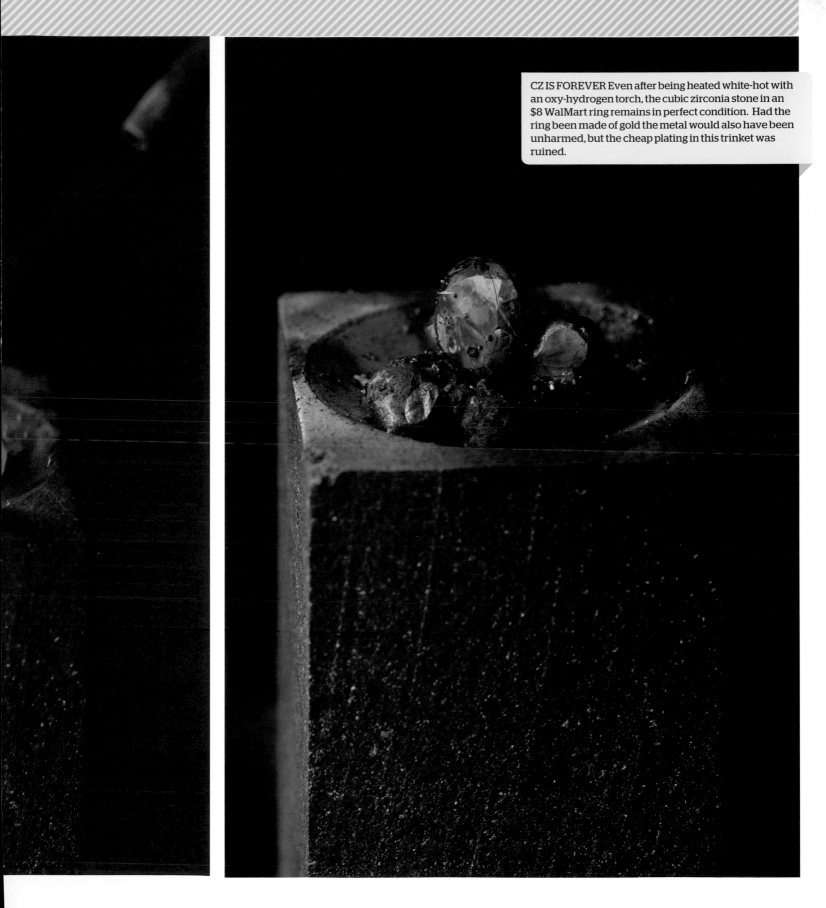

CZ IS FOREVER Even after being heated white-hot with an oxy-hydrogen torch, the cubic zirconia stone in an $8 WalMart ring remains in perfect condition. Had the ring been made of gold the metal would also have been unharmed, but the cheap plating in this trinket was ruined.

Spark Fun

An elemental recipe for creating great shower of sparks

METALS CAN BE CLASSIFIED by hardness, malleability and conductivity. One quality you won't find listed in the reference books is sparkiness.

A delicate balance between flammability and hardness determines which metals spark. For example, magnesium is a famously flammable metal, but grinding it produces no sparks because the energy needed to cut chips from the soft metal is not enough to heat them to their ignition point.

Although iron is much less flammable, it's so hard that separating chips of it heats them to the point that they catch fire and burn brightly as they fly off. The true champions of sparkiness, however, are the lanthanides—the elements from lanthanum (57) to lutetium (71). They are even more flammable than magnesium yet also hard enough to generate large amounts of heat when they are ground.

Lighter "flints" aren't made of flint but rather a mix of lanthanides, with iron added to tame the excessive sparkiness of the other elements. Lose the iron, and you have the pinnacle of sparkiness: Mischmetal (German for "mixed metal") contains lanthanum and cerium, with smaller amounts of other lanthanides. Blocks of it are often used in movie special effects. For example, a scene in which a car blows a tire and drags on its rims needs a stream of sparks coming off the wheel, so a block of mischmetal is strapped on for spectacular effect.

Ironically, modern aluminum wheels are nonsparking, so the only place this still happens at all is Hollywood.

HOT SHOWER
When a grinder is applied to mischmetal—a combination of various elements—it produces copious quantities of big sparks.

"A delicate balance between flammability and hardness determines which metals spark."

MAKE SPARKS FLY This campfire starter is basically just a giant stick of lighter "flint" (actually
mischmetal). Using either the built-in scraping blade or
a file you can produce great big sparks on demand.

How I Did It

Blocks of mischmetal can sometimes be purchased on eBay or from unitednuclear.com. If you don't find one at these websites, special effects suppliers in your area may carry them. Making sparks with such a block couldn't be easier: Just grind, file, or sand it vigorously. Remember to always wear eye protection when grinding.

To get the photo of a block being dragged behind a car, I built a clever hinged device that fit into the trailer hitch of my car. It allowed me to raise and lower a block of mischmetal attached to the hitch using a rope that led back to the driver's seat, so I could start and stop the sparks any time while driving.

While driving home from my shop I kept the block raised, but when I found myself being aggressively tailgated it was a struggle not to lower the block for what I immediately realized was its highest and best possible use: Scaring the crap out of tailgaters.

TAILGATERS BEWARE I towed a block of mischmetal along the road, turning my minivan into a fire-shooting menace.

HELL ON WHEELS A bicycle fitted with a sandpaper ring and a block of mischmetal creates flying sparks. What kid wouldn't want one of these?

Sweet Science

A glass of instant Kool-Aid requires eight ounces of water and a surprising amount of innovation

WHEN YOU THINK OF TECHNOLOGY, you probably think of computers and jet engines and such. But there are other feats of engineering that are equally sophisticated, just less obviously so. Instant Kool-Aid, for example.

There are two fundamental problems in creating a small tablet that quickly turns a glass of water into a fruity drink. The first is finding a way to disperse the ingredients without forcing an impatient customer to stir them. The solution is sodium bicarbonate and citric acid, a mixture perhaps best known as the plop-plop, fizz-fizz of Alka-Seltzer. These chemicals react with water to form carbon dioxide gas, breaking up the pill, dispersing its contents, and leaving behind just a small amount of sodium citrate, a harmless substance found in citrus fruits.

The second problem—making the drink sweet enough—is more difficult. It would take far too much ordinary sugar to fit in a small pill. This part of the problem was solved in 1879 with the chemist Constantin Fahlberg's discovery of the first artificial super-sweetener, saccharin. Instant Kool-Aid (introduced by Kraft as Kool-Aid Fun Fizz in 2011) incorporates a more recently developed sugar substitute, aspartame, better known as NutraSweet. Aspartame is 200 times as sweet by weight as cane sugar and

BIG PINK: A sugar-based tablet like the one we made [above, right] would have to be many times the size of an aspartame-based instant Kool-Aid tablet (above, left) to make the drink equally sweet.

"Aspartame, better known as NutraSweet, is

200 times as sweet by weight as cane sugar."

represents a very high level of sophistication in chemical engineering; it was discovered by a chemist who was assembling proteins into a polypeptide while trying to design a new treatment for ulcers. All the sweet needed for eight ounces of Kool-Aid fits in a pinch of aspartame.

The drink innovation I'd really like to see, though, still remains tantalizingly out of reach: a pill that heats a cup of coffee. Self-heating coffee does exist, but only in bulky containers that keep the chemicals that do the heating separate from the coffee you drink. Sadly for coffee lovers, no one has yet figured out how to make a mixture that fits in a pill and wouldn't poison you if you tried to drink it.

TINY BUBBLES There's no real reason for including these pictures of a fizzing pill dropping other than that they are pretty.

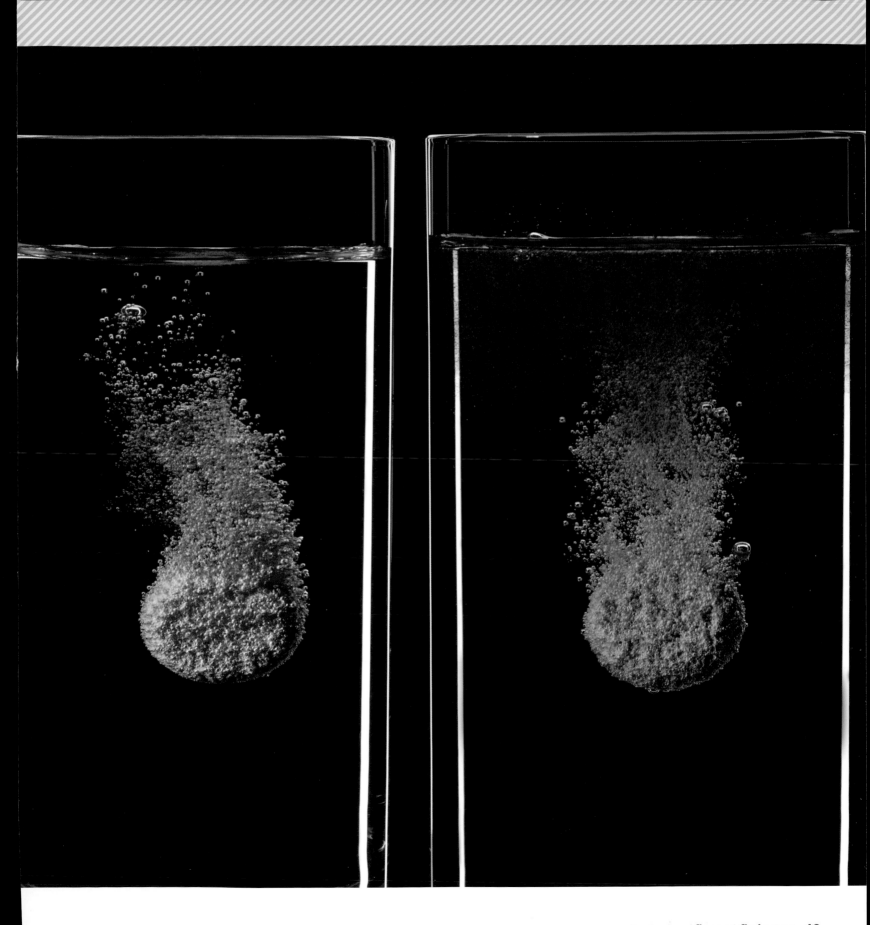

Tough Break

How science helps bike thieves

STRONG, HARD, TOUGH. These sound like different ways of saying the same thing, perhaps describing a really good suitcase. But when applied to the physical properties of materials, each of these words has a very specific technical meaning that distinguishes it from the others. And those definitions explain why it's so difficult to make a bicycle lock that can foil thieves.

Imagine a chain made of diamond. It would be impossible to cut with any hacksaw, but it could easily be defeated with a brick. Diamond is hard, but also brittle, much like glass. What diamond lacks is toughness: the ability to absorb energy without failing. A chain made of rubber would be far tougher in the technical sense. You could pound it for hours with a brick without breaking it. But it would not be hard enough to resist cutting.

It's actually not possible to make the hardest material also very tough. Harder materials can't yield to absorb energy and then spring back the way softer ones can. Their hardness means it requires less energy to break them. Take steel, for example. Bike-lock cables are made of very tough steel. You can't easily break a cable by hammering it. But compared with the high-carbon steel used to make tools, it's very soft, so it can be cut with a hacksaw.

Good locks are case-hardened, which means they have a hard but brittle outer layer protecting a tougher, softer inner core. But savvy thieves know that even good locks have a fatal flaw: Nearly any material, including steel, becomes less flexible when it's very cold. Although it doesn't lose any tensile strength—defined as how much force is needed to break it—the loss of flexibility makes it less tough.

When cooled to −13°F with canned air spray (actually the compressed chemical difluoroethane), even very tough locks become brittle enough to smash open with a hammer. So there isn't much that lock manufacturers can do to finally put an end to their long war with bike thieves. In the meantime, you might consider just taking your bike inside with you.

> **"It's not actually possible to make the hardest material also very tough."**

STONE-COLD LOCK This U-lock was specially treated to make it harder to break. But after cooling it with canned air, it cracked with just a few blows.

How I Did It

There is controversy about how well this technique actually works for stealing bicycles, and with how many different brands and models of lock. It definitely worked with the relatively cheap ones I got from WalMart, but more expensive ones may (or may not) be more resistant.

I used pretty much an entire can of canned air to get one lock cold enough, then gave it five or six really good whacks with the hammer before it shattered. The proof that this method is actually doing something is that when I tried the same amount of whacking (or even much, much more whacking) without first cooling the lock, it did not break.

One thing to be aware of is that they now put a bittering agent into canned air to discourage people from trying to get high off it. After emptying a few cans into my shop, I noticed a bitter taste in my mouth. It's odd, there isn't so much a smell in the air, it's more like a taste that comes from the air.

COLD FRONT If you hold a can of canned air upside down, it comes out as a liquid that creates a very strong cooling effect when it evaporates.

FATAL BLOW The cylindrical housing of this lock is made of forged steel, possibly case-hardened. It should be ductile enough to bend and dent when hammered hard enough, but when cooled with canned air, it cracks instead.

HARD VS. TOUGH At room temperature, a rubber ball is soft but tough and resists hammer blows. When cooled with liquid nitrogen to –321°F it's rock hard and cracks easily.

Bogus Bullion

How to make cheap fake gold

LAST SEPTEMBER, a New York City gold dealer spent $72,000 on his worst nightmare: fake gold bars. The four 10-ounce counterfeits came with all the features of authentic ingots, including serial numbers. That's pretty scary when you consider how many people own gold—or think they do.

I've been a fake-gold fan ever since author Damien Lewis wrote me into his 2007 spy thriller, Cobra Gold. My supposed experience making fake gold was pure fiction, yet I'm still treated as a source on the matter. I decided it's time to call my own bluff and make some real bogus bullion.

Instead of a 10-ounce ingot, I cast a two-kilogram (4.4-pound) fake the size of a Twinkie cake. A Twinkie heavier than four pounds? Yes, gold is dense, much denser even than lead. Good forgeries must have the right weight, and there is only one element as dense as gold that's neither radioactive nor expensive. That's tungsten, which can cost less than $50 a pound.

To fabricate a convincing fake, crooks could pour molten gold around a tungsten core. The bar would have a near-perfect weight, and drilling shallow holes would reveal nothing but real gold. A two-kilogram bar made this way would cost about $15,000 and be "worth" about $110,000. Since I have to work within PopSci's modest budget, and I'm not a criminal, I settled for a fake costing about $200 in materials.

I encased a tungsten core in a lead-antimony alloy, which is roughly as hard as gold. That way it feels and sounds right if touched and knocked. I then covered the alloy with genuine gold leaf to give my bar its signature color and luster.

My fake wouldn't fool anyone for very long (a fingernail can scratch off the gold leaf), but it looks and feels remarkable, even next to my real 3.5-ounce solid-gold ingot. Or at least I think that one's real.

"To fabricate a convincing fake, crooks could pour molten gold around a tungsten core."

REAL OR FAKE? Lead bars coated in imitation gold leaf [top left, top right] seem like knockoffs next to real bullion [cylindrical, flat bars] and a fake covered in genuine gold leaf [center].

How I Did It

The hardest part of this project was making the graphite mold. Which is ironic because really there's no need for a sophisticated high-temperature graphite mold for casting lead, I just wanted to see if I could do it (and a graphite mold would be required to cast real gold, which has a much higher melting point). Since I don't have a fancy computer-controlled milling machine, I used a cross-slide mounted on a tilted rotary table to guide a rounded-end milling bit around the angled sides of the mold. Then I sanded the inside of the mold silky smooth with fine sandpaper (graphite is very soft and easy to work).

Normally an even harder stage would have been making a block of tungsten the right size to fit the mold. Tungsten is incredibly hard to cut and cannot be cast, so it is usually formed into its final shape by hot pressing from powdered form: Not something that can easily be done at home. I avoided this problem by making the mold the right size to fit a block of tungsten I already had.

To get the lead to flow all the way around the block, I had to heat up both the mold and the block of tungsten. Otherwise the lead would have cooled almost immediately upon hitting the block. Unfortunately I didn't have it quite hot enough and a small gap remained on the bottom side. Rather than start all over, I managed to fix the hole with a bit of extra lead and the very careful application of a blow torch.

Applying the gold leaf was easy, it's a standard technique done using "gold size," which is basically glue. The leaves of genuine gold are incredibly fragile, and are usually applied using a squirrel-hair brush and static electricity. I couldn't find my squirrel-hair brush, so I just set the block down onto the sheets of leaf to get them stuck on.

Finally I used standard steel letter stamps to add the fake markings, and an arbor press to push a letter seal into the top surface of the bar.

IT'S A WRAP Gold leaf is so thin it spreads on more like paint than metal.

Impractical Knowledge

The Invisible Sea

You can't see them, but other gases can collect in places where we expect there to be air

ON JULY 2, 2007, Scott Showalter climbed into a manure pit on his Virginia farm to clear a blocked pipe. Moments later, he fainted and died. An employee of his went in to save him but was quickly overcome as well. One by one, his two daughters and wife followed, only to die trying to save the people who went before them. The culprit was invisible, odorless, tasteless methane gas, which had collected in the enclosed pit.

The sensation you normally experience when suffocating—the painful struggle to breathe—comes not from lack of oxygen but from a buildup of CO_2 in the bloodstream. That's why an asphyxiant gas is incredibly treacherous: If you are clearing your lungs by breathing it in and out, the CO_2 buildup does not occur, and you don't notice anything wrong before passing out.

It's difficult to visualize how gases could pool in a confined area. But it turns out their behavior is very much like that of liquids, so much so that you can create layers of gases of different densities—and even float boats on them.

Sulfur hexafluoride (SF_6), typically used as a spark suppressant in high-voltage equipment, is an invisible gas that's nearly five times as dense as air, making it ideal for a demonstration of this phenomenon. After allowing SF_6 to slowly fill an aquarium, I slid off the cover, leaving a pool of invisible gas that would stay put for several minutes. When I set a lightweight tinfoil boat on top of the tank, it floated on what appeared to be nothing at all.

It's really a magical sight, and a reminder that things are not always what they seem. A septic tank, a well or a mine can look perfectly normal, but if the air in it has been displaced by something more sinister, the only sign you'll get is the loss of consciousness. You will never know what hit you.

"It turns out the behavior of gas is very much like that of liquids."

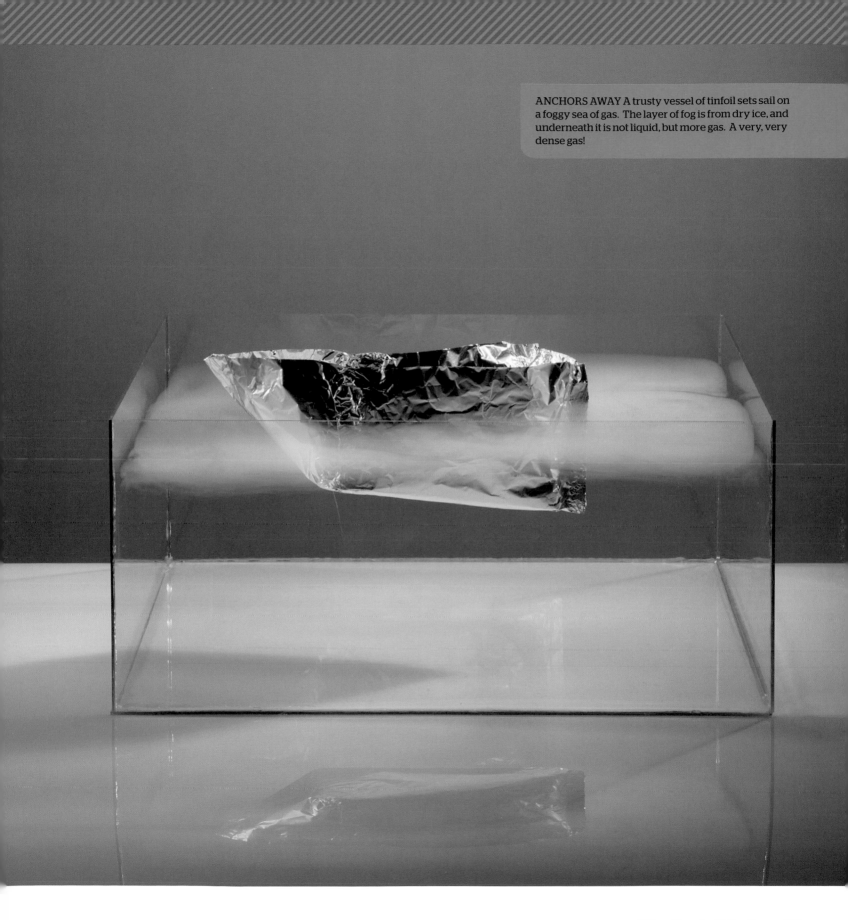

ANCHORS AWAY A trusty vessel of tinfoil sets sail on a foggy sea of gas. The layer of fog is from dry ice, and underneath it is not liquid, but more gas. A very, very dense gas!

How I Did It

The main problem with doing this demonstration is that sulfur hexafluoride is very expensive. A filled cylinder can easily set you back $500, and will only contain enough to do the demonstration three or four times. Is it worth it? Hey, whatever floats your boat.

To fill the aquarium, I pressed a sheet of lightweight plastic down into it with a gas hose underneath the plastic, then released the gas slowly and waited until the plastic had been raised up nearly all the way to the top (it doesn't need to be sealed airtight, just loosely pressed in). Then I slowly slid the plastic away, trying not to disturb the gas any more than necessary, and set the prepared boat on the invisible sea.

Sulfur hexafluoride can also be used as a sort of anti-helium: If you breathe it in, it makes your voice super-low, like James Earl Jones. Then you need to stand on your head for a while to get the gas to flow out of your lungs, where it tends to settle due to its high density.

I keep a large Mylar party balloon filled with the stuff to amaze visitors: It is delightful to drop it and hear the dull thud as the balloon slams into the ground. People think it's filled with liquid or something, it's hard to believe it really is just gas in there. (Though of course if it were actually filled with water, it would be far too heavy to even lift: Sulfur hexafluoride is heavy, but it's not liquid-heavy by a long shot.)

KEEPING A LID ON IT Sulfur hexafluoride is much denser than air, but also very expensive. A cover keeps air currents from sloshing the precious gas out as it fills the aquarium.

FOGHAT Dry ice fog is heavier than air, but not as heavy as sulfur hexafluoride, which allows it to settle just above the dense gas, making the invisible sea visible.

Change Agent

When there's not enough heat for a chemical reaction, add a catalyst

THE COPPER EARRING you see here had already been glowing bright orange for half an hour when we took the photograph. There is no flame under it, no electric current through it. Underneath is a pool of volatile and highly flammable acetone, but the liquid is not on fire. So where is the heat coming from?

Acetone vapor and oxygen from the air are combining and releasing heat on the surface of the copper, at a much lower temperature than acetone normally burns (but still hot enough to make the earring glow). The copper provides a sort of backdoor that overcomes the resistance (called the activation energy) that normally prevents acetone from reacting, except at higher temperatures.

Copper enables the reaction, but it is not consumed by it. (The earring can keep doing this indefinitely without being used up.) That property defines what it means to be a catalyst.

The most familiar examples of cata-lysts are the catalytic converters in cars, which finish the incomplete combustion of gasoline using platinum or palladium. Catalysts are important long before the gas makes it to the car, too. Platinum and rhenium are used to "reform" crude oil: to rearrange hydrocarbons into the specific molecules that make up gasoline.

Catalysts greatly reduce the energy, time and complexity of equipment required to do the reforming, and thus make the process far more efficient. In this sense, catalysis can be very green, which seems ironic when describing anything about the petroleum industry. In fact, catalysts can reduce energy use in a wide range of other large-scale chemical manufacturing processes as well.

I like the commonly used slogan "a catalyst for change" because it's a rare example of a phrase that is scientifically perfect. It describes an organization that makes things happen while continuing unchanged in its mission. That, in simple terms, is exactly what catalysts do.

"Copper enables the reaction, but it is not consumed by it; the definition of what it means to be a catalyst."

BUTTERFLY EFFECT
Copper facilitates a reaction between oxygen and acetone, causing an earring to continuously glow red-hot even without an external heat source.

How I Did It

This is a surprisingly satisfying demonstration for some reason. It seems to work well with any sort of real copper jewelry. Just make sure it's not varnished (sand or burn off any coatings). It's a bit touchy getting the earring hot enough just before lowering it into the pool of acetone, without setting the acetone on fire. Which brings us to the real issue: Acetone and its vapors are tremendously flammable. If you do this demonstration there is absolutely no question that the vapors will at some point catch fire. Expect it, prepare for it, don't be surprised when it happens, just cover the jar with something non-flammable to snuff out the flames. Never do this demonstration with anything flammable nearby, and store the rest of the acetone well away and tightly closed.

We had no problem getting a good photograph, because once set up, the earring would sit there for ten or twenty minutes on end, glowing consistently and brightly. Wafting air around the top of the container let us create patterns of light and dark on the earring, which made for a more interesting photograph. Left alone, it just glows uniformly.

THE FAMILY JEWELS Nearly any kind of reasonably pure copper jewelry seems to work for this demonstration, as long as it's not fake painted metal. Often a lacquer coating has to be burned off first to expose the raw copper surface.

REAL DANGER ALERT

Acetone vapors are extremely explosive. Using a blow torch near acetone is inherently dangerous.

A HEALTHY GLOW To start the catalytic reaction, the earring must be heated red-hot with a blowtorch. And this must be done very close to highly flammable acetone vapors. This is one experiment guaranteed to produce unexpected fireballs from time to time. Expect them.

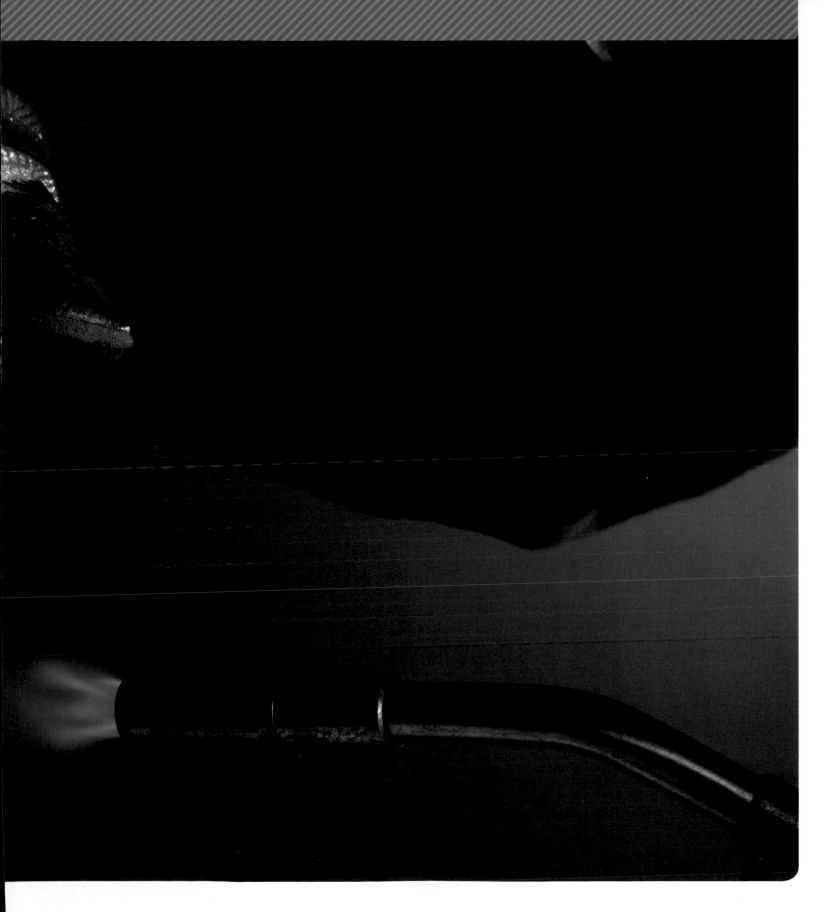

Hacking Light

Make groovy glowing designs with a little drain cleaner and a light stick

ONCE A PIECE OF MAGIC technology becomes so common that you can buy it at the gas station, people start taking it for granted. That happened with light sticks sometime in the 1990s. But with a little creativity, diagonal cutters and Drano, you can reveal—and control—that old black, green, red and blue magic inside.

When you bend a light stick, you break open a glass ampule inside. Diphenyl oxalate in the ampule reacts with hydrogen peroxide in the surrounding solution to form peroxyacid ester, a high-energy chemical compound. The color depends on a phosphorescent dye: When a high-energy ester molecule meets a dye molecule, the ester molecule decomposes and transfers its energy to the dye molecule, which emits that energy as a photon of visible light.

Despite the fact that glowing chemicals are usually associated with danger, light sticks are actually fairly benign. If you follow some precautions, you can open them up and experiment. For example, if you lower the pH of the solution by adding sodium hydroxide, also known as crystal Drano, ester molecules form at a faster rate, making the solution brighter (and using it up faster).

I used this effect to add a starry pattern to a psychedelic painting I made by pouring oxalate, peroxide and dye components onto a glass plate. The chemistry is the same as activating a light stick in the ordinary way, but this version just seems more . . . magical.

> **"When a high-energy ester molecule meets a dye molecule, the result is a photon of visible light."**

How I Did It

I bought several wholesale bags full of light sticks from an online supplier, cut them open, and poured the contents into various plastic bottles, giving myself a ready supply of different colors of light to paint with. In principle you could buy the chemicals directly in bulk form, but I found it to be a lot easier to get them out of glow sticks, which are relatively cheap and easy to get.

Cutting open a light stick isn't that hard, but the plastic is tough: I used diagonal cutters to nip the tops off. This is a very messy demonstration, and the dyes stain everything they come in contact with.

Avoid getting light-stick chemicals on your skin, and be very careful to keep them out of your eyes. The glass ampule inside could cut you just like any other broken glass, so wear gloves.

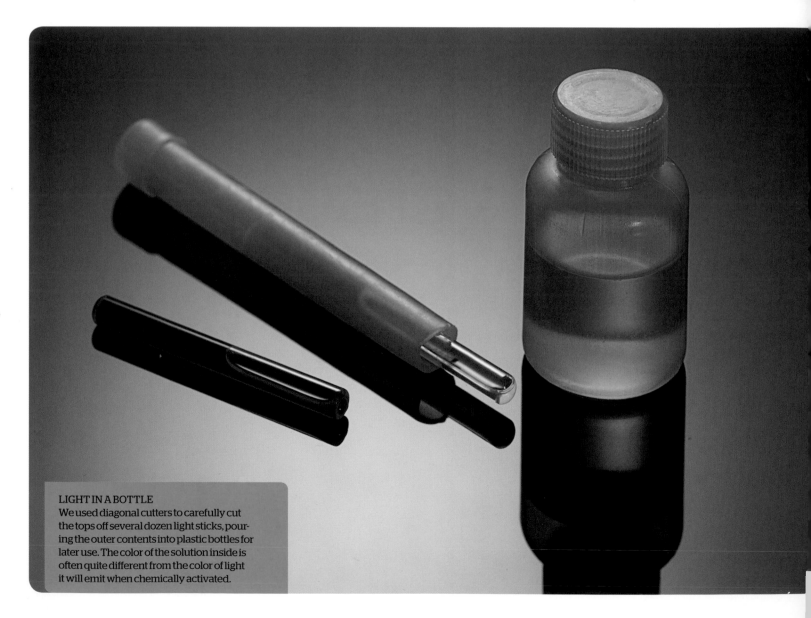

A DASH OF COLOR Swirls of light form when diphenyl oxalate is dripped slowly into a solution of hydrogen peroxide.

LIGHT IN A BOTTLE
We used diagonal cutters to carefully cut the tops off several dozen light sticks, pouring the outer contents into plastic bottles for later use. The color of the solution inside is often quite different from the color of light it will emit when chemically activated.

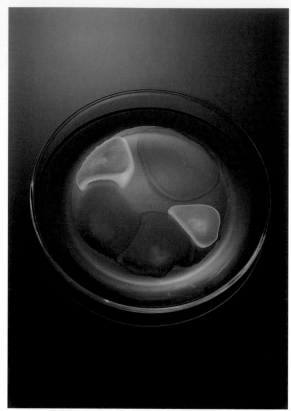

SUNNY SIDE UP
No, these aren't fried Smurf eggs. In order to create these glowing patches, small amounts of colored Diphenyl oxalate solution were poured in several locations onto a thin layer of hydrogen peroxide solution spread out on a glass plate. After those patches had spread out, different colors of the solution were poured into the center of each patch, creating concentric rings of psychedelic color. Grains of sodium hydroxide (Drano) were then sprinkled onto the plate, which creates bright sparkles as each grain accelerates the glow reaching around it.

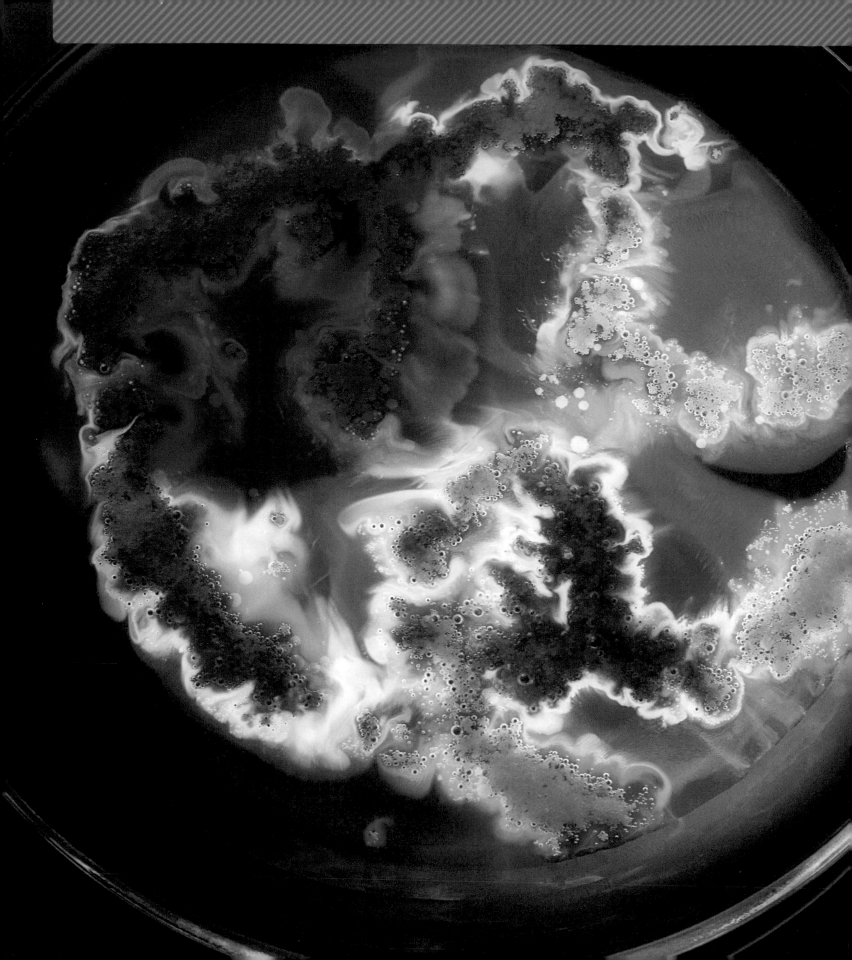

MAKE SURE BOTH SIDES CLOSE

DIY X-ray Photos

With some film and a saltshaker, you can take radioactive pictures

EVERYONE KNOWS light exposes film, but other forms of radiation do as well—a fact you can use to take pictures in some pretty unusual ways. It's also how radioactivity was first discovered.

In 1896, French physicist Henri Becquerel stored some x-ray film in a drawer along with a uranium rock. He suspected that uranium might emit strange rays when exposed to sunlight, but this sample had been kept entirely in the dark, so he was surprised to find, on developing it, that the mineral had exposed the film anyway. This discovery won him a Nobel Prize.

It's not hard to repeat Becquerel's experience at home with standard film. I took apart a 10-pack of Fujifilm ISO 3000 instant film and wrapped each piece in tinfoil. This must be done in absolute darkness because 3,000-speed film is extremely sensitive. (I sacrificed the first pack practicing in the light.)

Next I set a big, flat butterfly-shaped earring directly on top of the wrapped film. I suspended the most radioactive thing I have, a small radium puck from an old classroom set, several inches above the earring. This allowed the radiation to shine through it and onto the film, exposing it right through the foil wrapper. Then I developed the film by pulling it through the rollers of an old Polaroid camera (once again, in complete darkness).

This exposure took about 36 hours, determined by trial and lots of error. If you're willing to wait longer, less-radioactive sources work too, even common salt-substitute. Yes, sodium-free salt (potassium chloride) is sufficiently radioactive (from the isotope potassium-40) that after several months, a saltshaker-full will form an image on film. Provided you don't forget and eat the radioactive source on your breakfast.

> **"Henri Bacquerel was surprised to find that uranium, kept entirely in the dark, had exposed the film."**

How I Did It

This demonstration works best with high-sensitivity Polaroid film, which is getting harder and harder to find. It may soon in fact be impossible to get at all. And of course you need something radioactive: Antique Fiestaware from eBay should work well. Patience is required: The exposure time typically needs to be several days.

WRAP IT UP Tinfoil is used to protect the instant film from light. This step must be done in total darkness.

 REAL DANGER ALERT

Stronger radiation sources such as radium watch hands, and any source that's flaking off fine particles, should be handled with care to minimize exposure and avoid contamination.

THE MOMENT OF TRUTH The film is developed the film by pulling it through the rollers in an ancient Polaroid camera.

POSITIVE PICTURE When Polaroid film is peeled apart you get a negative (on the left) and a print (on the right).

In the Pink

Pepto-Bismol tablets contain a surprisingly large amount of heavy metal

MOST MODERN MEDICINES are carefully synthesized organic molecules so potent that each pill contains only a few milligrams of the active ingredient. Pepto-Bismol is a fascinating exception, both because its active ingredient is bismuth, a heavy metal commonly used in shotgun pellets, and because there is a *lot* of it in each dose. So much, in fact, that I was able to extract a slug of bismuth metal from a pile of pink pills.

One two-pill dose of Pepto-Bismol contains 262 milligrams (more than a quarter of a gram) of bismuth subsalicylate, and about 60% of that weight is bismuth. It's not just bits of ground-up bismuth metal, though. The bismuth is combined chemically with salicylate, an organic molecule. To get bismuth metal, you have to reduce it chemically, the way iron ore has to be reduced to make iron metal.

I tried reducing Pepto-Bismol by heating it with charcoal, the same method used to reduce iron ore, but that didn't work very well. All I got was crumbly slag. Fortunately, I found a better procedure with an assist from science-experiment website *thechemlife.com*, which recommends isolating the bismuth by reaction with aluminum in an acid solution. This way requires only muriatic acid (found in the paint department at my local hardware store) and aluminum foil (found in my kitchen).

The demonstration took a while: I had to grind, dissolve, filter, precipitate, and filter the stuff again. Not unlike the digestion process Pepto aids, you start with nice colorful morsels and end up with dark crud. For logistical reasons, I had not actually tried the method before we set up the photo shoot, so it was quite a joy when I first saw beads of liquid metal form as I heated the crud, telling me that we had not wasted an entire day on a wild bismuth chase.

"I was able to extract a slug of bismuth metal from a pile of pink pills."

How I Did It

This has got to hold the record for most useless demonstration ever, though it is fun. If you actually want bismuth metal, buy it from eBay at a thousandth the cost of extracting it from Pepto-Bismol!

4. After a very slow filtering step, I had a clear pink solution containing dissolved bismuth ions.

5. I dipped aluminum foil into the solution, turning it black. The acid dissolves the aluminum, which reacts with the bismuth ions. The ions precipitate out as bismuth metal particles.

3. I dissolved the pills in a solution of six parts water to one part concentrated muriatic acid.

6. After filtering the liquid through a pillowcase, the powder left in the filter is the bismuth metal.

2. I ground up the pills with a mortar and pestle.

7. Heating with a plumbing torch is the moment of truth: Will the black powder melt into shiny metal?

START HERE ⟶

1. I started with 180 generic Pepto-Bismol pills, which together contain about 24 grams of bismuth.

8. Success! A few grams of actual solid metal with the characteristic iridescent sheen of bismuth.

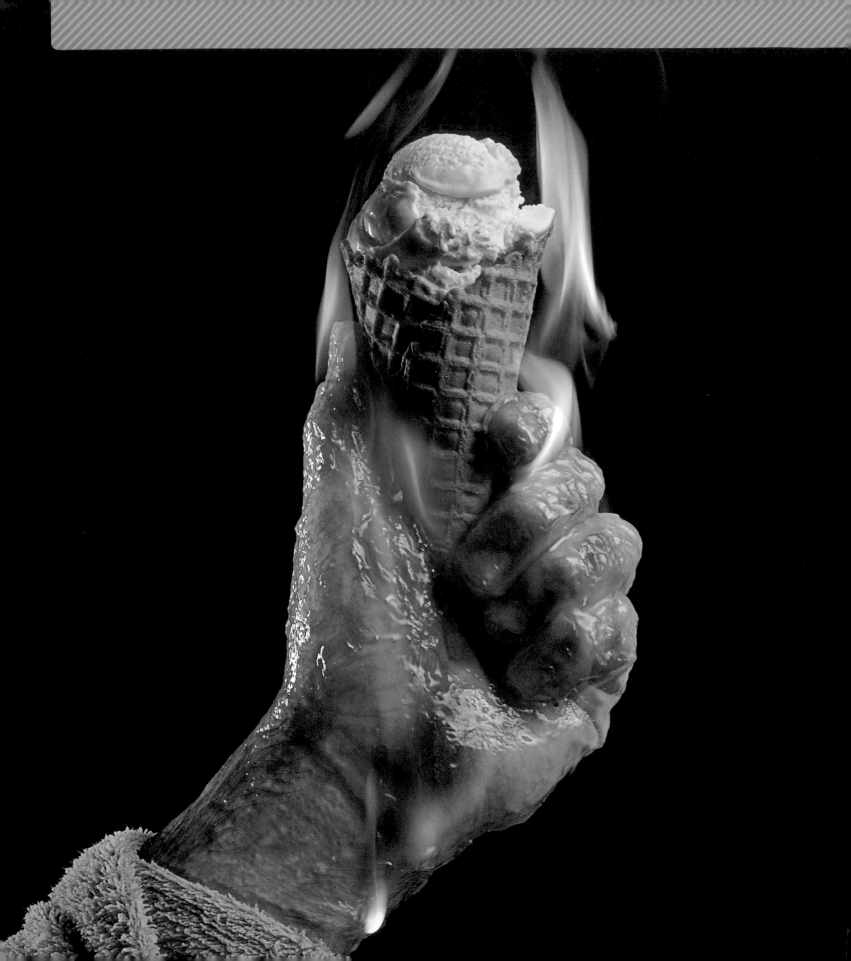

In Which
the Author
Endangers
His Hand

Cool Hand Theo

A layer of bubbles protects the flesh from liquid nitrogen, though only for a split second

WHEN I FIRST SAW this photograph of a man's hand submerged in liquid nitrogen at somewhere below –320°F, my immediate thought was, "That guy must be crazy! One second in that stuff, and you're shopping for new skin!" My shock was tempered only slightly by the fact that it was my hand, and we'd taken the picture just a minute earlier.

I hadn't realized that my hand was quite so deep into the liquid. Amazingly, I barely felt the cold at all. My skin didn't get hurt for the same reason that water droplets dance on a hot skillet. An insulating layer of steam forms almost instantly between the water and the metal, keeping the droplets relatively cool as they float for several seconds without actually touching the hot surface. To liquid nitrogen, flesh is like that skillet—a surface hundreds of degrees above its boiling point. So the moment my hand

touched the liquid, it created a protective layer of evaporated nitrogen gas, just as the skillet created a layer of steam. That gave me just enough time to put my hand in and pull it out again. Any longer than that, and frostbite would have set in.

The phenomenon is called the Leidenfrost effect (after Johann Gottlob Leidenfrost, the doctor who first studied it in 1756). I'd known about it for years, but when it came time to test it in real life, I have to admit that I used my left hand, the one I'd miss less.

I drew the line at another classic example of the effect. According to the books, it's possible to stick a damp finger directly into molten lead without getting burned, if you do it fast enough. After some consideration, and remembering the times I've been burned by molten lead, I decided that it probably wouldn't make a very good picture anyway.

> **"My skin didn't get hurt for the same reason that water droplets dance on a hot skillet."**

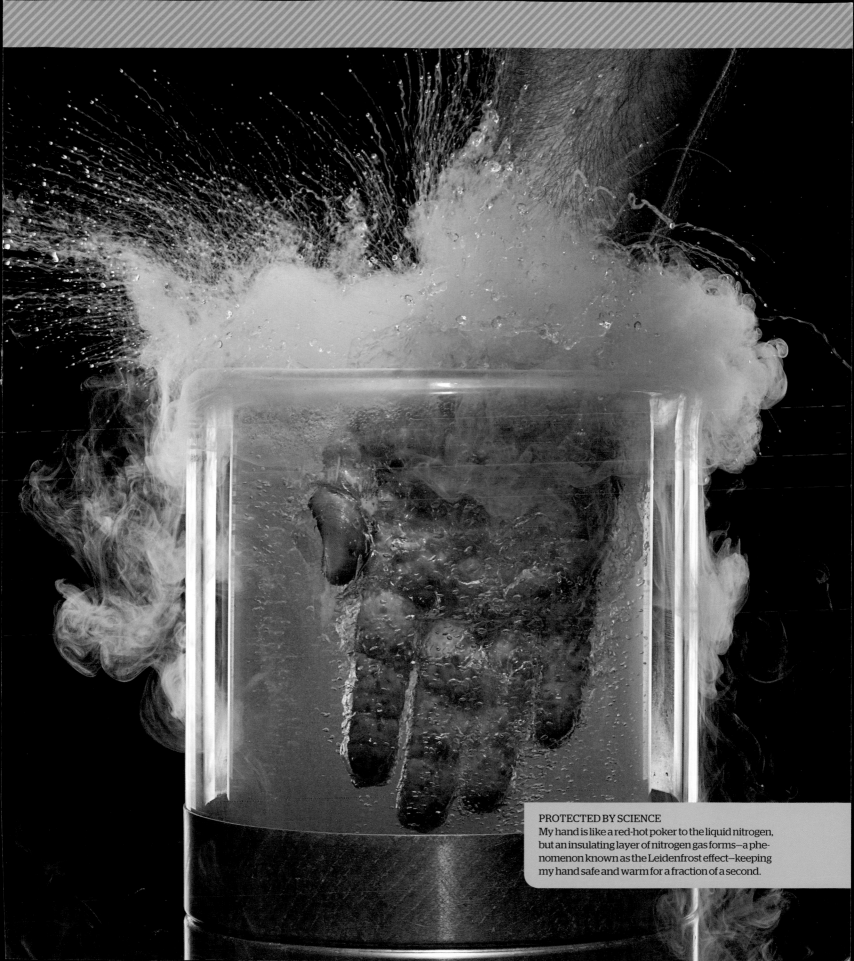

PROTECTED BY SCIENCE
My hand is like a red-hot poker to the liquid nitrogen, but an insulating layer of nitrogen gas forms—a phenomenon known as the Leidenfrost effect—keeping my hand safe and warm for a fraction of a second.

How I Did It

This is a very easy demonstration to do using a nice wide-mouth Dewar flask and plenty of liquid nitrogen. I was of course tentative at first, but soon realized that I could leave my hand in long enough to hit the bottom and get a good photograph. After that it's really quite fun: A cool and refreshing experience.

It's actually my eyes that would NOT be fine if, while hastily plunging my hand in an out of the flask, I were to splash some drops into them. Liquid nitrogen is not a completely harmless substance by any means.

DANCING DROPLETS Small droplets of water on the surface of a hot skillet actually last longer before evaporating than ones on a slightly less hot surface. When the skillet is above a certain temperature, the water on the bottom of the droplet evaporates fast enough to create a layer of steam that levitates the drop above the surface, insulating it temporarily from the heat of the pan.

 REAL DANGER ALERT

Do not try this. If liquid nitrogen soaks into your clothes, you will not be protected by the Leidenfrost effect, and you can get frostbite very quickly. If it gets into your eyes, it can blind you.

Poison Light

One of the nastiest substances on Earth creates a beautiful glow

BEING A MAD SCIENTIST can be a thankless job, but every once in a while you get a chance to shine—literally. I recently had that opportunity when working with a TV show to film one of the most beautiful of all chemical phenomena, the cold luminosity of white phosphorus.

White phosphorus, an uncommon form of the element, is terrible stuff. It's incredibly toxic, catches fire spontaneously in air, and is frequently used in the production of methamphetamine. There are few companies lining up to sell it, and the producers of the television show couldn't find any sane scientist able to get it and do the demonstration for them. I, however, keep some out back behind my house in an army-surplus ammunition case.

With my trusty Icelandic chemist friend Tryggvi at my side calculating the proportions to make sure we'd survive, I mixed up a solution of white phosphorus in toluene (paint thinner) and spread it on my hand (with latex gloves on—I may be mad, but I'm not crazy). The effect was something I'd only read and dreamed about: fleeting, flickering, flowing sheets of cold light all over my hand that rippled as I blew on them.

As the toluene evaporated, the white phosphorus did what its name implies—it phosphoresced, reacting with oxygen from the air to produce a luminous gas fractions of an inch above the surface of my hand. This reaction is not just risky to create, it's also difficult to photograph. We used the best low-light camera available to get it to show up so well.

To the dark-adapted human eye, though, it looks bright, and frankly rather alarming, especially when I took the gloves off and realized that toluene goes through latex. I didn't plan to have white phosphorus on my skin (thankfully, a tiny, harmless amount), but I have now seen my own bare hand flickering green with poison light, and it was every bit as satisfying as I thought it would be.

> **"White phosphorus, an uncommon form of the element, is terrible stuff.**
>
> **It's incredibly toxic and catches fire spontaneously in air."**

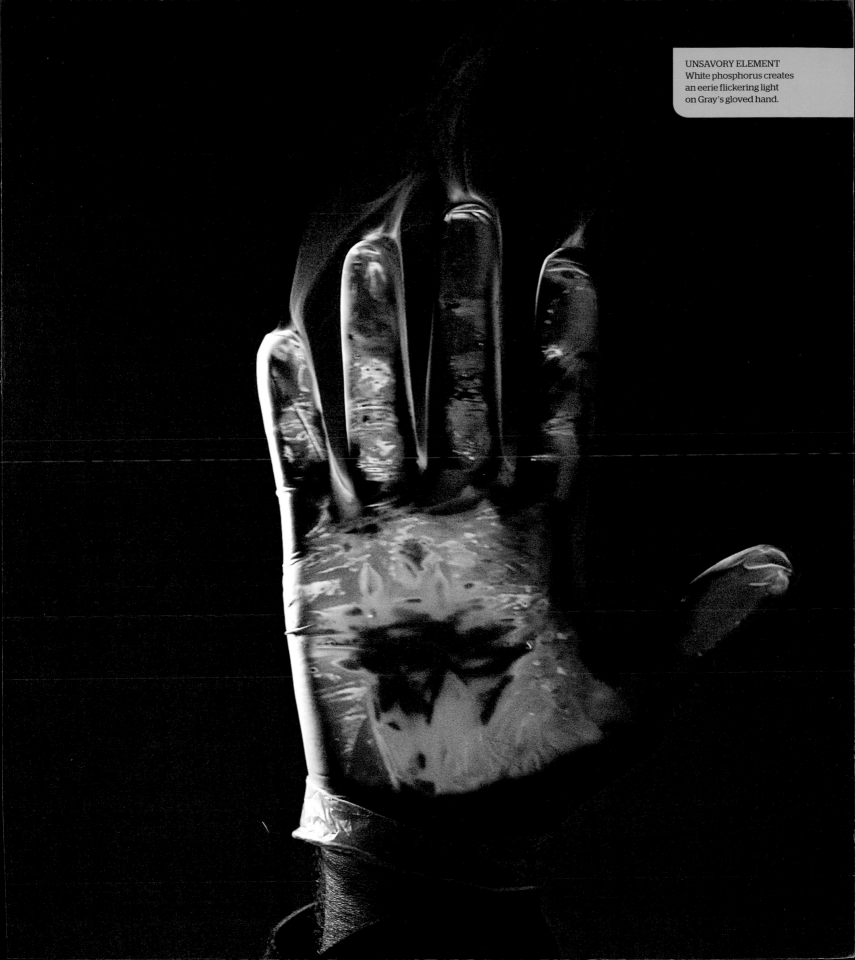

How I Did It

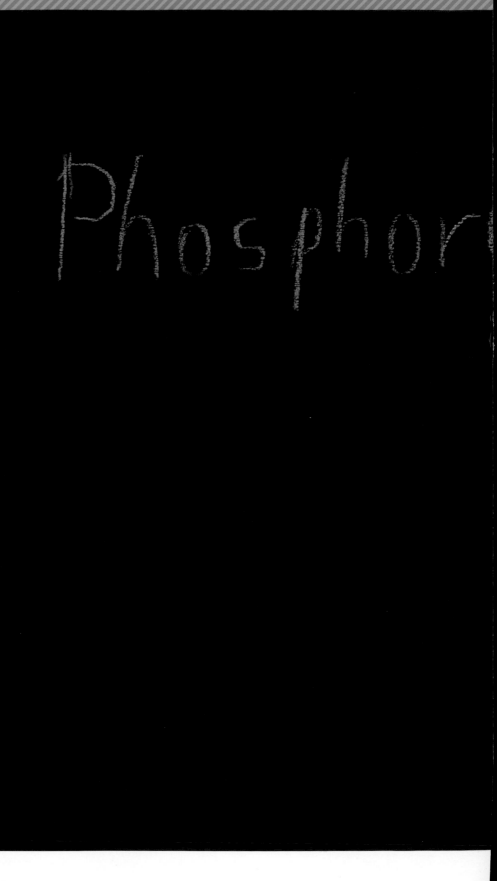

I cannot in good conscience recommend that anyone do this. I probably shouldn't have done it myself, but I just had so much white phosphorus, and NOVA asked so nicely (having been told no by the research university they approached first). The fact is that unless you have some already, you probably won't be able to get any white phosphorus anyway: It is not possible to buy it from any commercial sources, and its possession is not legal in many states.

My first step in doing the demonstration was to ask my friend Triggvi, a practicing professional chemist, to help. He suggested (and brought along) several different solvents to try dissolving the white phosphorus in. Most of them worked fine, and all of them ended up working their way through the two layers of nitrile and latex gloves we were wearing. Both Triggvi and I at one point had visibly glowing white phosphorus on our bare skin, which for some reason I found more alarming than he did. (He had done the math and determined that the total amount we could be exposed to, even in the worst case, was well below a toxic level. He trusts his math.)

It was a surprise to both of us that on more than one occasion, one of our gloves caught fire. This seemed to happen when the solvent evaporated and left a tiny amount of white phosphorus residue on the glove, which then ignited due to the warmth of our hands. This stuff is really sneaky.

The glowing crayon effect (seen in the photo to the right) was achieved by simply drawing on a piece of wood with a small chip of solid white phosphorus held in tweezers. We then used a hair dryer to heat up the wood and make the writing glow brighter.

REAL DANGER ALERT

Do not try this. Even a tiny amount of white phosphorus can be highly toxic if ingested, and larger amounts burn skin on contact.

MARK MY WORD
Letters written with a piece
of white phosphorus quickly
burn into a wood surface.

All AGLOW The red glow of this smoke is kind of fake. The glowing letters are real and the smoke is real, coming from the white phosphorus, but the red glow was created by an iPhone screen held just out of sight behind the wooden plank. Yes, there's an app for that (it lets you make your whole screen any color you like, for casting small amounts of colored light on your subject).

WHITE HOT WORDS A hair dryer blowing warm air onto white phosphorus letters greatly increases their brightness, and can even set them on fire. (The red glow is from the heating coils of the hair dryer: These photos were taken in very low light so even that dim glow is enough to contribute.)

Hot Tip

Trust in science allows for a quick dip of the finger in molten lead

PREVIOUSLY, I STUCK MY HAND in super-cold liquid nitrogen. My skin survived that demonstration [see p.84], but I wimped out on a related experiment at the opposite extreme: dipping my finger into molten lead. That's because the only time I've ever burned myself badly enough to need a doctor was while casting a lead plaque as a kid.

But life is too short to cower in the dark, afraid of a little molten metal. I knew that the antidote to my fear was science; I trusted the Leidenfrost effect to keep me safe. When my finger hits molten lead that's hot enough, moisture from my finger should be vaporized instantly, creating an insulating layer of steam that should protect me for a fraction of a second. This is a mirror image of what happened when I put my hand in liquid nitrogen, where the heat from my hand was hot enough to instantly vaporize the nitrogen, similarly creating an insulating layer of gas.

The "hot enough" part is key. If the metal is just barely molten, not enough steam is created, and some of the lead may solidify onto the finger, where it would rapidly transfer enough heat to cause a serious burn. So I had to get the temperature well beyond that level, test the metal with a hot dog, and then go for it.

In place of lead, I used nontoxic plumbing solder, which has a melting point of around 400°F. When the temperature got up to 500°, I inserted my pinkie finger—and didn't feel a thing. I managed to come away unscathed even though my finger was in the liquid past the knuckle long enough for me to splash around a bit. OK, according to the high-speed video we shot, it was only about one sixth of a second, but it was long enough to cure my childhood fear of molten lead for good.

> "When my finger hits molten lead that's hot enough, moisture from my finger should be vaporized instantly, creating an insulating layer of steam. The 'hot enough' part is key."

CYCLE OF LIFE In an attempt to add some pathos to the photos, we put a defective tin soldier into the pot.

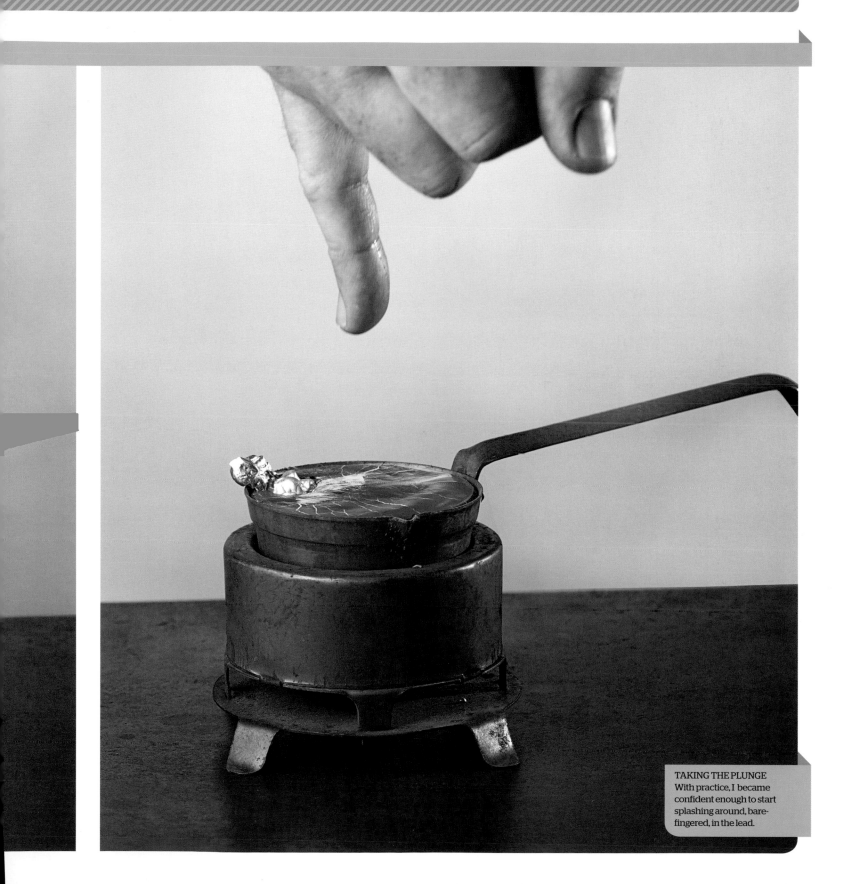

TAKING THE PLUNGE
With practice, I became
confident enough to start
splashing around, bare-
fingered, in the lead.

Saving My Skin

Any scientist can tell you how people catch fire in the movies. Only one sets himself ablaze

THERE ARE A FEW PERKS to my job as a mad scientist, and one of them, as I recently learned, is being able to tell my colleagues that I can't attend their terribly important meeting because I'm going to set my hand on fire.

In the movies, people on fire stumble out of burning buildings all the time. If you look closely, however, you'll notice that they are almost always fully dressed, and that they tend to keep moving. These are two important factors that make the stunt much easier.

To function and avoid injury while on fire, you need to put something between yourself and the flames. But you can't coat yourself with plain water because it just runs off. So stuntpeople use clothes containing super-absorbent polymer fibers, which keeps the water in place (it's pretty much the same material used in diapers). A layer of clothes treated this way will keep them cool for quite a while, but they have to continue to move forward so the breeze keeps the flames out of their uncovered face.

If a scene requires showing bare skin on fire, stuntpeople use a special fire-protective gel containing water, which can be applied in a smooth, clear layer that is nearly invisible, especially when the action is moving fast and there's a lot of fire to distract the viewer. To show you what it looks like up close, I covered my hand in the gel and then painted on some thinned-down contact cement, which produces a very nice opaque yellow flame when lit.

Because we needed my hand to be perfectly still for the camera, I couldn't use any movement to help stay cool, and my hand started to feel pretty warm after just a few seconds—but not before we got photographic proof that my meeting excuse was for real!

"In the movies, people on fire stumble out of burning buildings all the time."

THE HUMAN TORCH
Sodium polyacrylate polymer in the gel on my hand protects it from the fire.

HOT HEAD With a waterlogged hat for protection and the heat from the fire going up and away, my assistant, Taylor Walker, could keep his head comfortably on fire for more than a minute.

How I Did It

The key to fire effects is an assistant with a bucket of water or wet towel nearby. Fire protective gel protects you from fire long enough to be able to respond in a calm and controlled manner when that protection starts to wear off. Remember, calm and controlled: Just because your hand is on fire and it's starting to get warm is no excuse to panic.

The hardest part of this demonstration was getting the gel nice and smooth on my hand. It's much easier if you're trying to make a piece of clothing burn, because the gel can be hidden under the cloth.

FIRE AND ICE(CREAM) (Above) Gray and the photography crew went through quite a bit of smoked ice cream to get this shot. [Left] The brand of fire protective gel used is mainly intended to protect nearby surfaces during welding. Contact cement thinned with naptha makes very nice yellow fire with just the right amount of particulate smoke.

REAL DANGER ALERT

Do not try this—people who play with fire get burned. This stunt requires professional training and preparation, and still there are plenty of things that can go wrong.

Making Stuff

Getting the Lead Out

Seemingly harmless children's toys have long been made from highly hazardous materials

AMONG THE MOST strictly enforced consumer-protection laws are those banning lead in toys. Lead is an insidious poison: It's slow-acting and results not in immediately noticeable effects like rashes but in behavioral problems and a slightly lowered IQ. Even a very small amount of it is harmful. Yet a few decades ago, a lot of the most popular playthings were made from solid lead, including tin soldiers.

Considering all the lead toys produced in those days, tin soldiers sound pretty benign. "Tin" is a something of a misnomer, though. The soldiers were not made primarily of tin but of a lead-tin alloy containing 60 to 75 percent lead, with the rest being mostly tin and antimony. Sometimes they were cast from "hard lead," a group of alloys typically found in bullets, which contain nearly 95 percent lead with just a bit of antimony for hardness.

Children didn't just play with these little chunks of neurotoxin; they often cast them in their own kitchens, using kits that came with a melting pot, a ladle, some sticks of lead alloy and a selection of soldier molds. After casting, kids filed them smooth (spreading lead dust all around). Then they decorated their armies with a variety of paints, most of which were lead-based.

"A few decades ago, a lot of the most popular playthings were made from solid lead, including tin soldiers."

BAD CASTING Making tin soldiers from molten lead alloy was a popular activity for children as recently as the 1960s. These soldiers were made from lead-free plumbing solder.

Safety standards, thankfully, have progressed significantly since then. At today's standard, 100 parts per million or less, just one of those old soldiers contains enough lead to render several million toys unfit for sale in the U.S. Although such safety requirements have no doubt helped reduce the number of lead-poisoning cases, they may not be stringent enough. Unlike with most toxic substances, there is no limit below which lead is known to be harmless. As more evidence of lead's deleterious effects on the brain accumulates, it would not be surprising to see the 100-ppm standard lowered further. If you really want to play with tin soldiers safely, you'll have to find some vintage silicone rubber molds and cast them from lead-free plumbing solder, as I did.

How I Did It

Creating tin soldiers is a fun hobby, and there's no reason a careful child under adult supervision couldn't cast such figures at home. Old but perfectly usable molds and whole kits are available on eBay, and if you use pure tin for casting there isn't a problem with lead contamination. If you use a tin-lead alloy (which are commonly used and commonly available) you do want to worry about not getting it in pots also used for food, and not getting filings all over the kitchen.

Tin and lead can be melted on a kitchen stove in a stainless steel or cast iron (not aluminum) pot: You don't need any special equipment. Sometimes the first few castings come out missing limbs and other details: That gets better as the mold heats up and the metal is able to flow in further before cooling.

REAL DANGER ALERT

Whether it contains lead or not, molten metal can burn and spatter out of molds unexpectedly. Always use proper safety equipment.

Cold, Hard Facts

How to cast solid, if fleeting, shapes in mercury: Just add a lot of liquid nitrogen

WHAT YOU CONSIDER SOLID, liquid or gas depends entirely on where you live. For example, men from cold, cold Mars might build their houses out of ice. Women from Venus, where the average temperature is about 870°F, could bathe in liquid zinc.

We think mercury is a liquid metal, but it's all relative. At one temperature, the mercury atoms arrange themselves into a solid crystal; at another, they flow freely around each other as a liquid. Children from Pluto (like mine, for example) could happily cast their toy soldiers out of mercury, because on that frigid planet it is a solid, malleable metal a lot like tin. Here on temperate Earth, you need a stove to cast tin but a tank of liquid nitrogen to make mercury figurines.

At liquid-nitrogen temperature, about –320°F, mercury acts like any other metal: You can hammer it, file it, saw it. (It won't shatter like other liquid-nitrogen-frozen items because there's not enough moisture inside.) Watching it solidify is exactly like watching tin harden from a molten state. As the atoms go from liquid to solid crystal form, you see the surface pucker. And because mercury, like most metals, shrinks when it solidifies, you see the surface sink in areas, forming a patchwork characteristic of cast metal.

The fun of making frozen mercury trinkets is another reason to lament the fact that this marvelous metal is also an insidious poison that must be handled carefully and never spilled. Schools have been evacuated because of one broken mercury thermometer, and mercury in the environment, particularly in fish, is a major public-health concern. Which is, of course, why I made this cute little mercury fish.

"At one temperature, the mercury atoms arrange themselves into solid crystal; at another, they flow freely around each other as a liquid."

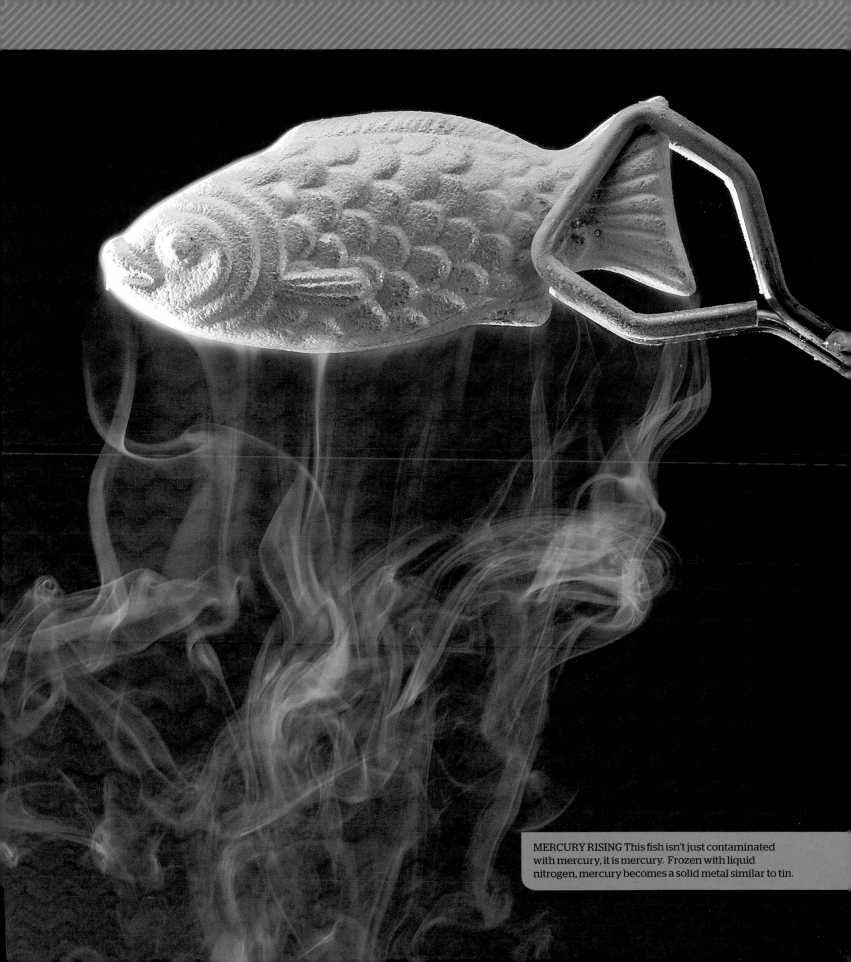

MERCURY RISING This fish isn't just contaminated with mercury, it is mercury. Frozen with liquid nitrogen, mercury becomes a solid metal similar to tin.

NICE CASTING Room-temperature mercury poured into a cold cornbread mold.

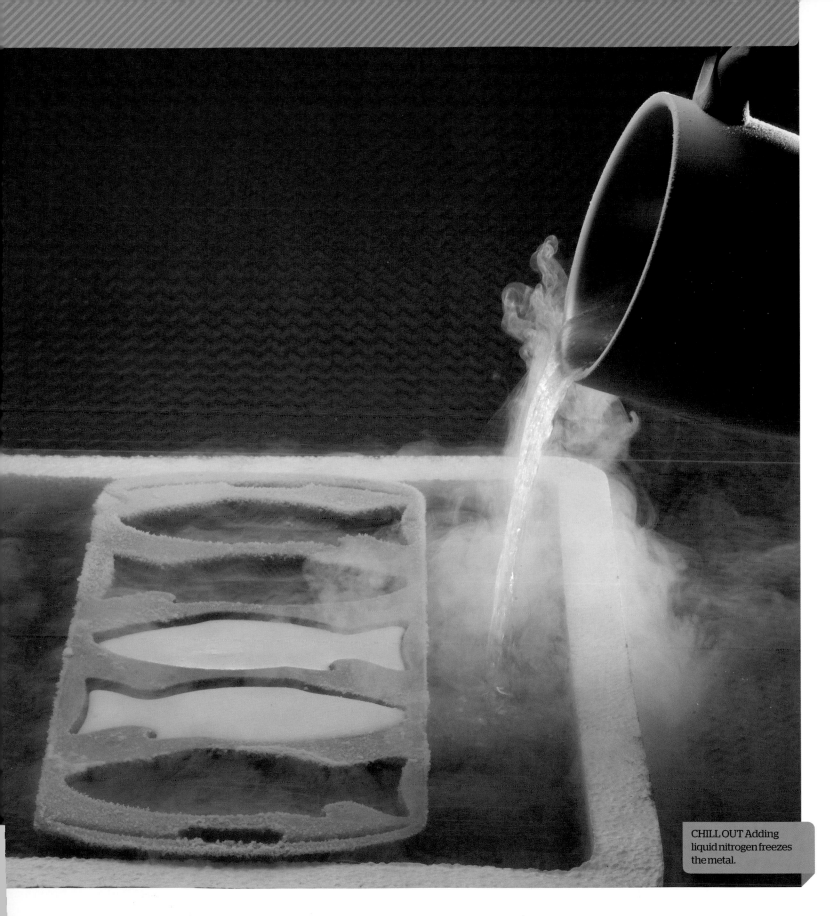

CHILL OUT Adding liquid nitrogen freezes the metal.

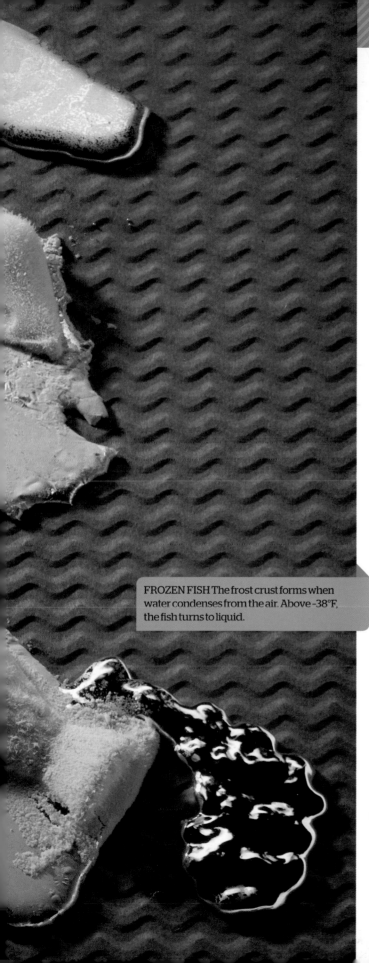

FROZEN FISH The frost crust forms when water condenses from the air. Above –38°F, the fish turns to liquid.

How I Did It

This demonstration is one that comes in the category of not really a good idea unless you have a very good reason for wanting to actually do it (for example, you want to take pictures for a book you're working on). It is not difficult to cast and freeze several pounds of mercury in a mold, but you're working with several pounds of mercury, and even a few grams of mercury spilled can be a major hassle to clean up.

I set a cornbread mold in the shape of a fish in about half an inch of liquid nitrogen poured into a cut-down Styrofoam cooler. Then I poured room-temperature liquid mercury into the mold and let it solidify. What struck me immediately was how similar watching the mercury solidify was to watching molten tin solidify. In both cases, near the end you see crystals forming in the center, and a small depression or cavity forms as the metal contracts on cooling.

 REAL DANGER ALERT

Don't handle mercury. It is toxic, and even minor spills can be dangerous and very expensive to clean up.

ETCHING APPARATUS
Clamps from a battery charger run about 10 amps between an electrode and the metal being machined. The flowing saltwater acts as a conductor. The drill press simply moves the electrode.

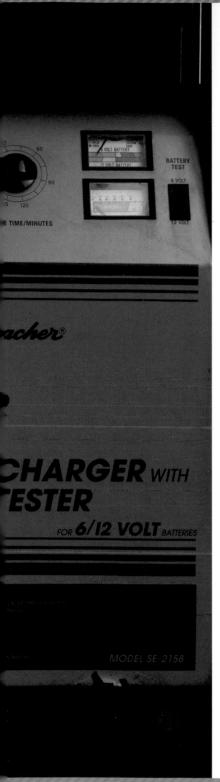

Power Sculpting

Carve steel with saltwater, electricity and a tin earring

I REMEMBER SEEING a demonstration of a seemingly magic process at an engineering open house decades ago, in which a soft metal bit carved detailed shapes into far harder metals. It's called electrochemical machining (ECM), and it's so simple in principle that you can do it at home with a drill press, a battery charger and a pump for a garden fountain.

ECM is basically electroplating in reverse. In electroplating, you start with a solution of dissolved metal ions and run an electric current through the liquid between a positive electrode and the object you want to plate (the negative side). The ions deposit themselves as solid metal onto the surface of the object.

In ECM you start with plain water (lightly salted to make it conductive) and run the current in reverse, so you're turning solid metal on the piece you're machining into dissolved ions in the saltwater, wearing it away a tiny bit at a time. The shape of the electrode determines the pattern that results.

Because it's the electric current doing the work, the electrode never touches the other piece. And it works equally well no matter how hard the metal is. I used a cheap soft-tin earring as an electrode to machine a simple shape into a washer made of hardened steel. After about 15 minutes, the earring was still good as new, but the steel of the washer was eaten away to less than half its thickness.

This process is used industrially to create extremely delicate, detailed shapes in very hard metals. Since there is zero force exerted on the part being machined, it's possible to make fine shapes that would break if you tried to cut them with a milling machine.

Without the precise current control of those commercial systems, my home setup produced a disappointingly blurry copy of the earring's shape. But it's still amazing that in a contest between a tin earring and hardened steel, the car ring won.

"ECM is a process used industrially to create extremely delicate, detailed shapes in very hard metals."

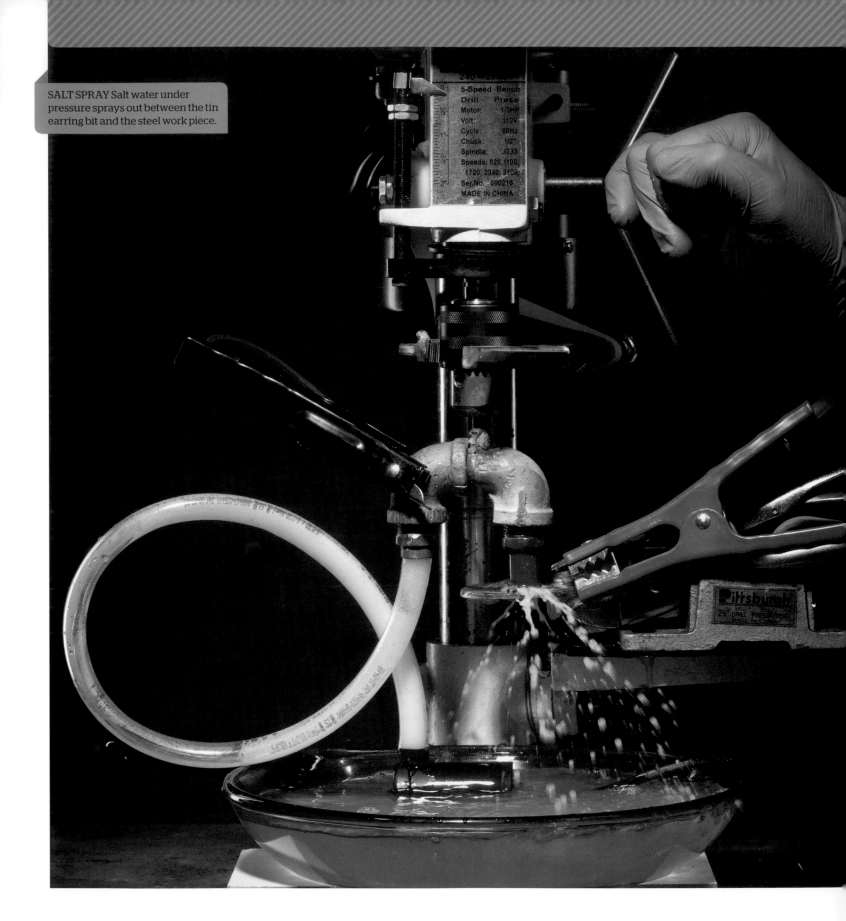

SALT SPRAY Salt water under pressure sprays out between the tin earring bit and the steel work piece.

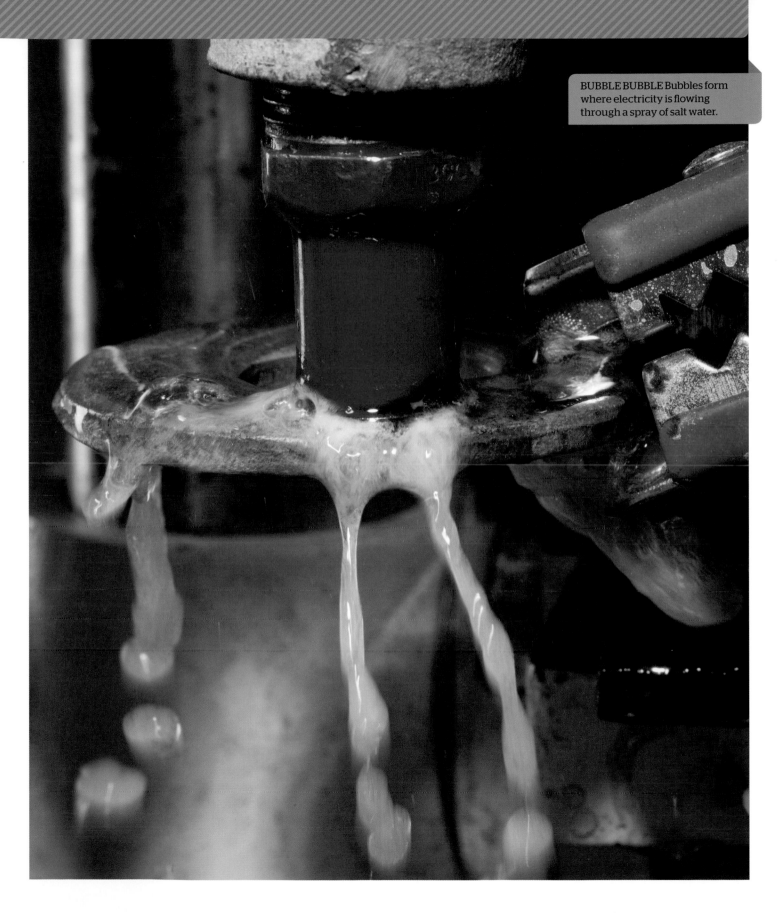

BUBBLE BUBBLE Bubbles form where electricity is flowing through a spray of salt water.

FIRST IMPRESSIONS This effort at electrochemical machining resulted in a fairly pathetic, blurred dent. But done right, this technique can produce exquisitely detailed shapes.

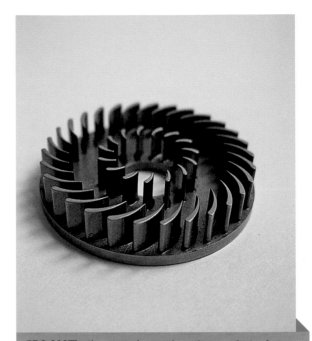

PRO-AM The tin peace-sign earring acts as an electrode, etching away the metal in the hardened steel washer [left]. The imperfect results are due to the difficulty of manually maintaining an exact thousandths-of-an-inch distance between the two. Commercial electrochemically machined pieces, like this microturbine for a water pump, use sophisticated electronics to monitor the current flow and carve precise pieces [above].

REAL DANGER ALERT

When your hands are wet with saltwater, even 12 volts is potentially dangerous. Wear rubber gloves.

Practical
Fire

Hold Your

SCREEN TEST A fine-mesh kitchen sieve with a candle inside simulates a Davy miner's safety lamp. An explosive mixture of propane gas and air is blown in from the outside. If the mesh is fine enough, the fire will stop at the screen even as the explosive gas flows through it.

Fire

A wire screen is all it takes to prevent dangerous gases from exploding

IF YOU WERE A COAL MINER in the early 1800s, the light you used was an open-flame oil lamp—even though mines were sometimes filled with "fire-damp," a volatile mixture of air and methane gas. Explosions were inevitable, and at times threw bodies from mine shafts like grapeshot from a cannon. Humphry Davy became a national hero when, in 1815, he found a remedy: Surround the lamp flame with mosquito screen.

Davy, one of the world's first professional scientists, solved the problem by systematically studying what happened when gases burned. He started with the observation that gas flames would not travel down long, thin metal tubes because the metal draws heat from the flame, lowering the temperature of the gas below the ignition point.

He tried making the tubes thinner and shorter, until he discovered that thin tubes needed to be only about as long as their diameter to prevent fire from traveling their length. The logical end point was fine metal mesh, which you could think of like thousands of very short tubes arranged in a grid.

As bizarre as it sounds (and I really didn't believe it until I saw it with my own eyes), you can blow an explosive mixture of gases through a fine wire mesh toward a candle flame and, when the gas explodes, the fire stops dead at the mesh—this despite the fact that the gas came right through the holes in the mesh and is just as explosive on the outside as it is on the inside.

Davy refused to patent his invention, preferring to bask in the glory of his role as the miner's savior. His lamp remained in use, and his name was a household word, right up until the invention of electric light. Today his lamp is all but forgotten, but his reputation as one of the first and greatest chemists lives on.

How I Did It

When attempting this demonstration, the most important thing is to make sure that there aren't any leaks around the mesh where flame can find its way through. Notice the small pile of sand at the base of the mesh in the photograph to the right: That's to seal off the bottom. I used an aluminum mosquito screen, and the setup is as you see it, no special tricks to get it to work.

In my experience propane works well, but not acetylene. Acetylene is just too flammable, and the flame passes easily through the wire mesh. Even with propone it will occasionally burn through, which means that good ventilation is a must to avoid blowing up the whole room.

The Davy lamp was at first considered a great boon, because it prevented explosions in mines when miners encountered pockets of explosive gases. Later people started to realize that mine operators were just using it as a crutch to avoid the better solution of making sure there weren't any pockets of explosive gases in their mines in the first place. Because, lamp or no lamp, if there's a pocket of explosive gas, there's a danger of it igniting one way or another.

FLAME OUT (Below) Without a wire screen to contain them, flammable gasses enthusiastically fill the air with flame. (Right) Mosquito netting and a kitchen strainer is all it takes to contain nature's fury.

☠ REAL DANGER ALERT

Anytime you're playing with explosive gases and open flames, the result will be dangerous blasts of fire. The Davy lamp does not work with all kinds of flammable mixtures, nor does it work under all conditions.

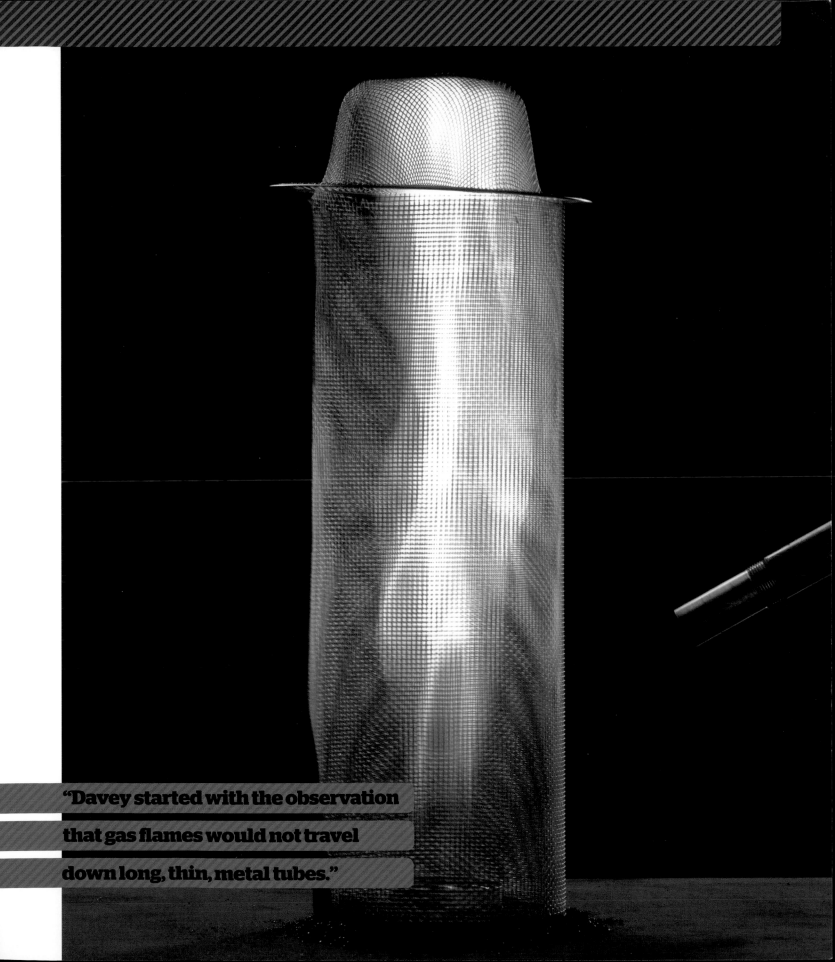

"Davey started with the observation that gas flames would not travel down long, thin, metal tubes."

Hot Under Pressure

How to start a fire with nothing more than compressed air

YOU'VE PROBABLY SEEN contestants on Survivor trying to make fire by rubbing sticks together or concentrating sunlight with their eyeglasses. But among preindustrial fire-starting methods, it's hard to beat the portable convenience of fire pistons, used in Southeast Asia since prehistoric times.

Almost all gases heat up when compressed. The harder and the faster the compression, the hotter the gas gets, hot enough even to ignite cotton wool or other flammable materials. Diesel engines work the same way: They have no spark plugs; instead the fuel/air mixture is ignited by compression as the cylinder closes up.

Perhaps most surprising is that this same principle also explains how many high explosives work. They are called "high" because their explosive reaction expands through a supersonic pressure wave that travels much faster than ordinary burning, making them far more powerful than low explosives like gunpowder. Each successive bit of material in a high explosive ignites when the pressure wave compresses and heats trapped microscopic bubbles of gas. When manufactured without bubbles, even extremely powerful high explosives can be impossible to detonate. Without gas to compress, there is no way for the detonation wave to heat up neighboring areas.

For example, ANFO (ammonium nitrate/fuel oil) explosive mixtures, commonly used in mining, don't always naturally contain enough trapped gas, and require a "sensitizer" to render them reliably explosive—often just a slurry containing hollow glass microspheres.

Some high explosives also create heat through the friction of microscopic crystals rubbing against each other, but in many cases the difference between bang and no bang is just hot air.

"It's hard to beat the portable convenience of fire pistons, used in Southeast Asia since prehistoric times."

PISTON PYRO A thick-walled acrylic tube with aluminum plunger (from $45; survivalschool.com) forms a demonstration fire piston, revealing a bit of cotton set alight by pressure alone.

How I Did It

It is possible to make fire pistons that work out of all kinds of materials, but this is a case where I decided to buy a commercial unit because there was one available that was exactly what I needed. So really it was just a question of getting the timing right to capture the flash of light in the photograph, which was a matter of luck and perseverance.

FANCY FIRE A beautifully made wood-and-metal fire piston (from $65; wildersol.com) designed for lighting campfires or other survivalist needs

Fire in the Hole

Chemicals can turn any tree stump into mock gunpowder

WHEN YOU NEED to remove a tree stump, you have several options. Sissies call a tree service. Tough guys loop a chain around the stump, hook it to the bumper of their truck, and find out which one is stronger. Others use gunpowder to blow them up, though this is not advisable in most jurisdictions (unless your cousin is the sheriff and you let him watch). But my favorite method is to convert the stump itself into gunpowder and then burn it up. That is the secret behind how chemical stump remover works.

You might think you could just light stumps on fire and let them burn until they disappear. But since they're underground, there's no source of oxygen to sustain the flame. Even with kerosene soaked into the wood, the part of the stump under the surface won't burn. Gunpowder, on the other hand, burns even inside a sealed space because it contains its own source of oxygen in the form of potassium nitrate, or KNO_3, better known as saltpeter. Get saltpeter into the stump, and it supplies oxygen to combust the wood.

Most common brands of chemical stump remover are nothing more than saltpeter. The instructions say to drill holes down into the stump, pour in the powder, and let it soak with water for up to a few months. This dissolves the saltpeter and distributes it throughout the stump. Then you soak the stump with kerosene and light it, causing it to burn all the way down to the roots with a fizzing, popping, purple-blue flame.

The stump's altered chemical composition—potassium nitrate combined with organic carbon to produce heat and gas—is similar to gunpowder. That explains the unusual flame. The burn is slower, though, taking minutes instead of milliseconds to complete.

It might be surprising to discover that you can buy the key ingredient in gunpowder at any garden center. But here's the kicker: The other two ingredients are readily available as well. If you want to find out what those are, read about my adventures making gunpowder (see page 146).

> **"My favorite method for removing a tree stump is to convert it into gunpowder and then burn it up."**

SLOW BURN
Wood treated with stump remover burns with a purple-blue flame at and below its surface.

How I Did It

This is a very simple demonstration, all you need is a log, a drill, and some commercial stump remover (read the label to be sure it's potassium nitrate). Drill a hole in the top of the wood, fill it with KNO_3, use some water to soak the KNO_3 into the wood, then let it dry out thoroughly before lighting it.

WOOD BE GONE Stump remover used in combination with kerosene can get rid of a stump pretty fast. Not as fast as dynamite, but a lot faster than rot.

REAL DANGER ALERT

Commercial stump remover is usually made of potassium nitrate plus a few impurities and stabilizers. Be sure to carefully read and follow the instructions on the label, because each brand is a little different.

FIRE IN THE HOLE Potassium nitrate stump remover melts and bubbles in the heat created by its own burning, sputtering up out of the holes meant to contain it.

Flash Bang

The hidden uses of everyday explosives

THE EXPLOSIVE C4, a favorite for everything from demolition to terrorism to action movies, is in fact one of the safest explosives. How can an explosive be safe? If it's hard to set off by accident. C4 is so stable that you can light it with a match (it burns but does not explode) or shoot it (it splatters but does not explode). To go bang, it requires a detonator that produces both heat and shock.

At the other end of the spectrum are mixtures that ignite simply from being scratched or knocked. There are obvious challenges in mixing, storing and handling these substances so that they explode only when intended, yet they're surprisingly common.

If matches, for instance, weren't friction-sensitive, they would be useless. Matches use potassium chlorate and red phosphorus, two chemicals that ignite when pushed into each other by friction. Strike-on-box matches isolate the phosphorus in the striking surface, while strike-anywhere matches keep the two chemicals in separate zones of the head.

Ammunition is another example. Gunpowder will not ignite from impact, so most cartridges contain a primer, usually a dab of lead styphnate, which turns the whack of the firing pin into an explosion that lights the gunpowder.

The most unexpected place I've found one of these contact explosives is in MagiCube camera flashbulbs, which were popular in the 1970s. They had absolutely no electrical components, not even a battery. Instead they contained four glass bulbs filled with flammable zirconium wool held in pure oxygen. Emerging from the sealed bulbs were thin metal tubes filled with a shock-sensitive pyrotechnic mixture. When you pressed the shutter button, it released a wire, which smacked the side of the ignition tube. The burning primer ignited the zirconium wool just like the primer ignites gunpowder, resulting in a flash of super-bright light—and giving a whole new meaning to the term "photo shoot."

FULL CARTRIDGE

"The most unexpected place I've found one of these contact explosives is in

MagiCube camera flashbulbs, which were popular in the 1970s."

How I Did It

This was a quite technically challenging photograph to get, which makes it all the more surprising that the picture you see on the previous page is from the very first attempt. We never got another one anywhere near as good.

I started by pulling the slug (bullet) off the end of a common 9mm center-fire pistol cartridge: With the cartridge in a vice, I grabbed the slug with pliers, then twisted and pulled to extract it. I poured the gunpowder onto some paper to save it. Using a fine-tooth hacksaw I very carefully sawed away a section of the cartridge down to the level where the percussion explosive primer was contained. To get the cross-section shot of the cartridge I then replaced the cutaway section of cartridge with scotch tape, poured the gunpowder back in, and set the slug on top.

To get the shot of the cartridge firing, I built an elaborate assembly of three articulated vices. One vice held a pair of vice grip pliers holding the empty cut-away cartridge (no gunpowder, no slug, just the primer). Another vice held a flat piece of brass that I was using as a leaf spring. And the third held another pair of vice grips that in turn held a partially disassembled center punch pointed at the cartridge. The brass leaf spring was set up resting against the back end of the center punch, such that when it was pulled back and let go, it would whack the center punch, just like the hammer of a pistol whacks the firing pin. The leaf spring was held back by a rod with a long string tied to it, so I could release it from a distance by pulling the string. A sound trigger connected to our camera captured just the right moment, first try.

Even though there was no gunpowder or slug in the cartridge, I had no idea how powerful the primer charge would be, so I was pretty nervous releasing it for the first time. The answer is that the primer is easily powerful enough to knock the cartridge out of even very tightly clamped vice grips: Eye protection was a good idea for this demonstration.

INNER WORKINGS With the cover removed you can see that this camera is able to trigger its flash with absolutely no electronic components. It uses mechanical and chemical components only.

Playing with Fire

Spontaneous combustion is easier than you think—if you know how to do it

THE FACTS ABOUT spontaneous combustion are easily lost. Mostly this is because spontaneous human combustion is a favorite among conspiracy-theorist types. Reports of people suddenly going up in flames tend to omit an essential detail, such as a lit cigarette. Yet as with many phony scientific concepts, the possibility is so intriguing that some people just want to believe.

The same is true of spontaneous nonhuman combustion. A video made the rounds on YouTube that showed a cotton ball catching fire after having been soaked in several tubes of Super Glue. How cool is that! Of course, I had to try it—again and again—with every kind of Super Glue and cotton ball I could find. It never worked.

There's no question that Super Glue gets really hot when mixed with cotton. The high surface area of the fibers causes the glue to harden very rapidly, releasing energy in the form of heat. Manufacturers warn about burns caused when Super Glue drips onto clothes, which has happened to me personally. The hot fabric gets stuck to your skin, and any attempt to pull it off just means that it also gets stuck to your fingers.

But there's a big difference between burning skin and an actual flame. As far as I can tell, the cotton ball in the video does not catch fire; a closer look at the footage suggests that the video was edited just before the blaze starts.

The beautiful thing about science is that for every fraudulent phenomenon, there's a real one that's even more extraordinary. Take, for example, a lesser-known form of spontaneous combustion that is genuine and easy to reproduce: spontaneous enema combustion (no, really).

To demonstrate it, I made a depression in a small pile of finely ground potassium permanganate, the chemical sold for recharging iron water filters. Then I squeezed the contents of a glycerin enema dispenser into the depression. Potassium permanganate is a powerful oxidizer and reacts with the carbon-hydrogen bonds in glycerin. A few seconds later, the mixture burst violently into flame. It worked every time—no mystery, no doubt about it.

> **"The beautiful thing about science is that for every fraudulent phenomenon, there's a real one that's even more extraordinary."**

PAN ON FIRE Lush purple flames and smoke, smelling vaguely of almonds, burst from a pile of potassium permanganate activated by a glycerin enema.

How I Did It

Potassium permanganate and a glycerin enema is just about the coolest way ever to create fire. It works absolutely reliably every single time, and the fire created is very hot and very purple. Why use a glycerin enema rather than just a bottle of glycerin? Because it's much funnier. And actually, the enema dispenser is legitimately a very convenient way to dispense just the right amount of glycerin in one squeeze.

I keep a jug of potassium permanganate and a few boxes of enemas handy at all times, because it is just too much fun to show visitors this quick and reliable demonstration.

TOOLS OF THE TRADE Potassium permanganate is sold in quart jugs in stores that have water purifier supplies. Glycerin is commonly available in many forms at any drugstore.

REAL DANGER ALERT

Super Glue on clothing can cause burns. Potassium permanganate and glycerin is a serious incendiary.

OUT OF THE FRYING PAN The fire and smoke from enema combustion is particularly attractive.

Impractical Fire

A Not-So-Sweet Surprise

Sugar seems harmless, but in powder form it can be a deadly explosive

IT'S ALL SWEETNESS and light until the sugar hits the fan. In 2008 the sweetener killed 14 people in Georgia—not from diabetes or heart disease, but in a violent explosion. Absence of regulation, ineffective enforcement, and lack of preparation for the potential danger led to the Imperial Sugar factory disaster, one of the worst industrial accidents of our time.

Any substance that burns at all can be made to burn bigger and faster by increasing its surface area, thus increasing its access to oxygen. Fine powders can have very large surface areas, making materials such as flour and sugar, which barely burn when in bulk form, explosive enough to demolish a factory.

To demonstrate the awesome power of confectioners' sugar, I blew a stream of it over a candle flame, creating a two-foot-long burst of fire. If the powder is instead dispersed in an enclosed space before ignition, it all burns in a fraction of a second. With nowhere for the pressure to go, it will cause a thundering explosion.

Few people truly appreciate this hazard. Fortunately, a renewed political interest in effective safety regulations has led to new enforcement and educational campaigns that should help keep powder-plagued workers safer.

But there's still a ways to go. A few years ago, I watched my neighbor auger his beans up into my spare grain bins. The air was thick with dust—and he was smoking a pipe. We didn't die that day, but seeing him with that glowing pipe in his mouth was one of the scariest experiences of my life.

THE SETUP Gray used a candle, a plastic bottle and fine confectioners' sugar. Regular granulated sugar did not work, but a turkey baster could substitute for the squeeze bottle.

"Any substance that burns at all can be made to burn bigger and faster by increasing its surface area, thus increasing its access to oxygen."

ON THE LEVEL Blowing too low made the candle flame go out; too high, and the sugar didn't catch. But when done just right, the room filled with heat, light and the pleasant smell of burned sugar.

 REAL DANGER ALERT

Never light flammable dust in a closed container: It really is explosive. Lit out in the open, the resulting fireball can singe or set on fire anything flammable in the area.

A Blast from My Past

While other kids played cops and robbers, our author made his own gunpowder

FOR AS LONG AS I can remember, I've loved gunpowder. One of my fondest childhood memories is pulling down volume G of the encyclopedia and seeing the formula for this magic substance for the first time. Saltpeter, sulfur and charcoal, listed with exact percentages! That was heady stuff for a kid who had been forced to rely on collecting match heads for flammable material. But where to get the ingredients? I settled on hitting up pharmacists, telling one that my mom had sent me out to get saltpeter for canning, and a different one that she'd sent me out for sulfur and I didn't know why (because I couldn't think of a better cover story).

What I didn't know is that all the ingredients for gunpowder are readily available side by side, no questions asked, in any garden center or home-improvement store. Charcoal is sold for grilling, and sulfur comes in bags that say "sulfur" in big letters (nice old ladies use it for dusting roses). But the key secret I never figured out back then is that most brands of stump remover are little more than pure saltpeter.

To make real gunpowder that actually goes bang, these ingredients must be ground together in a ball mill or stone rolling mill for hours, during which time there is a good chance that the powder will explode prematurely. (Seriously, don't try this at home unless you have the correct type of remotely operated mill.) Instead, I always ground the ingredients separately with a mortar and pestle and then mixed them gently without further grinding. This results in a powder that burns energetically but slowly: perfect, it turns out, for making sparkler cones.

I never got hurt, and with the kind of gunpowder I was making, common sense was enough to keep me in one piece. That wasn't the case for everything I used to experiment with, though. Sometimes I cringe when I think about all the times I was lucky not to blow my head off.

> **"What I didn't know is that all the ingredients for gunpowder are readily available side by side, no questions asked, in any garden center or home-improvement store."**

COURTING DANGER
My home-made sparkler cone shoots fire two feet in the air. My record as a kid was about five feet.

How I Did It

Crude non-explosive or minimally explosive gunpowder can be made like I describe here, using a stone or ceramic mortal and pestle to grind up the ingredients before mixing them. The material burns far more slowly than real gunpowder because the particles are much bigger and much less well-mixed. To create sparkler cones I just wrapped a bunch of layers of paper into a cone shape about 5 inches tall, filled it up with the powder, and taped the whole thing up with a lot of duct tape.

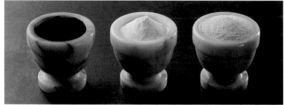

BEHIND THE BANG The ingredients for gunpowder [from left: charcoal, saltpeter and sulfur] come in a lovely variety of colors.

 ## REAL DANGER ALERT

Creating and igniting pyrotechnic mixtures of any kind, including gunpowder, is inherently dangerous and is illegal in some places. Harmless experimentation, especially by kids, can be taken very seriously by the authorities, so an adult *must* always be present and take full responsibility.

AIM, FIRE: A pile of gunpowder lit out in the open burns rapidly, but will not explode, because gunpowder, even real professionally made gunpowder, is not a high explosive.

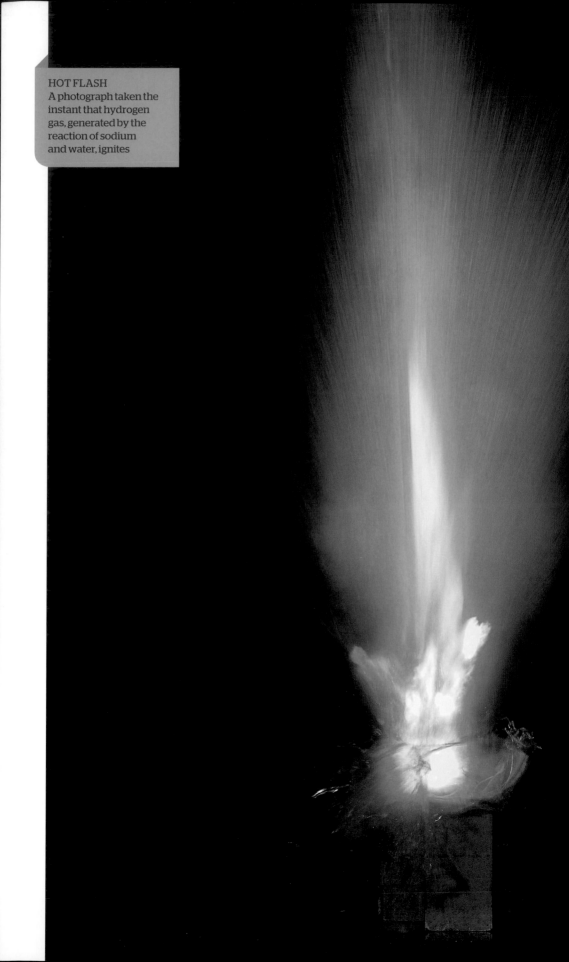

HOT FLASH
A photograph taken the instant that hydrogen gas, generated by the reaction of sodium and water, ignites

Big Bang Theory

Alkali metals react strongly with water—but which one creates the biggest explosion?

RUNNING DOWN THE far-left column of the periodic table, the readily available alkali metals—lithium, sodium, potassium, rubidium and cesium—all generate potentially explosive hydrogen gas when they touch water. The strength with which they react with H_2O goes up steadily in the order listed. Lithium just sizzles, whereas cesium explodes powerfully and instantly. You'd expect that to mean that cesium makes the biggest explosion, but it's not the case.

In fact, the British TV science show *Brainiac* demonstrated alkali-metal explosions a few years ago, and when the producers found that cesium failed to produce a giant explosion, they fixed the results using dynamite. (Which is a lousy thing for a science show to do—it made me feel bad for years because I could never replicate their demonstration.)

The power of the explosion depends on how quickly the metal gets hot enough to ignite the hydrogen gas that the reaction generates. Lithium typically never gets to the threshold of about 930°F needed to burn the gas, so there's no explosion. Potassium, rubidium and cesium react so fast that they immediately ignite the hydrogen, resulting in rapid burning but small explosions.

The sweet spot is sodium: It generates lots of hydrogen gas that builds up above the water, and then seconds later the gas ignites, producing a thunderous explosion that throws molten sodium balls in all directions. Good times if that's what you were looking for; permanent blindness if you weren't expecting it.

As the old saying goes, in theory there is no difference between theory and practice, but in practice there is. That's why sodium—the cheapest and least toxic alkali metal—is by far the most fun one to throw in a lake.

"All alkali metals generate potentially explosive hydrogen gas when they touch water."

How I Did It

Creating sodium explosions has been a favorite pastime of mine for years. The trick to getting these photographs was to control the process as much as possible, while dealing with the degree to which it really can't be controlled.

I started by building a small device with a metal cup (about the size of a shot glass) on a hinge with a rod holding it upright, and a block of wood holding the rod in place. When I pulled the block of wood away using a long string, the rod dropped down, allowing the cup to flip over, dropping the sodium into a bowl of cold water underneath it. (See photo of contraption at right.) Cold water works better than hot water, which tends to cause the reaction to happen too fast, resulting, surprisingly, in smaller explosions.

When you drop sodium into water, it first sits for some time fizzling as it generates hydrogen gas. Then, a completely unpredictable number of seconds later, that hydrogen explodes violently. There is absolutely no way to know how long it's going to take before exploding, and the explosion only lasts a few milliseconds. This makes it tricky to capture the light of the explosion: A standard camera shutter is not fast enough to open fully before the explosion is finished, even if you trigger it the instant the explosion starts.

My solution was to do the photography at night, and set up an optical trigger (which detects the light of the explosion) in such a way that it closed, rather than opened the shutter. After turning off all the lights, I opened the camera shutter, then pulled the string which dropped the sodium into the water. A few seconds later it exploded. A few milliseconds after that the optical trigger fired the flashes, which let you see the bowl and smoke just the right amount of time after the explosion had set them in motion, and a few milliseconds after that the shutter closed, capturing the shot.

BLOWIN' UP When dropped in water, potassium [right] catches fire instantly and throws molten material several yards, while cesium [left] creates a larger explosion. The reactions of these metals are not nearly as powerful as sodium explosions, however.

☠ REAL DANGER ALERT

Handling alkali metals is advanced chemistry—they're extraordinarily dangerous and can blind you with flaming balls of liquid metal that react with skin and eyes.

ALL BLOWN UP This photo was taken milliseconds after the hydrogen-air explosion was ignited. The combustion is so fast that no fire is still visible, only the results: water and red-hot sodium flying in all directions.

The Line of Fire

My brave volunteers get a flaming handful of bubbles

I **RECENTLY APPEARED** on a Japanese TV show and this was one of the demonstrations I performed. The setup was very simple: Get a bunch of minor Japanese celebrities to line up with their hands outstretched holding a line of bubbles and then light the bubbles on fire.

There was an easy part and a hard part to redoing it in Illinois. Flammable bubbles? Easy. We took soapy water and added a mix of butane and hydrogen, which reacts with the oxygen in the air to make for a good display. The more hydrogen, the faster it burns; about one-quarter hydrogen worked best.

The hard part was finding minor Japanese celebrities willing to have their hands set on fire. In the end I had to use employees of my software company instead, plus an assortment of my son's friends.

Why, you might ask, would anyone want to do this demonstration? Well, on the show there was one other factor I haven't mentioned yet. The poor guy at the end was secretly given a special mixture of pure hydrogen and oxygen gases, which exploded with a bang when the fire reached him. The shock that resulted was perfect for the weird world of Japanese TV, where practical jokes go way beyond what the lawyers would allow here. It was pretty funny to see the look on all their faces when they realized they'd just been had by a crazy American.

But exploding bubbles are actually quite dangerous. Get the mixture or quantity wrong, and injury is much more likely than with bubbles that just burn gently. So for this article, we didn't do that part. Fortunately, the lack of a bang at the end was more than made up for by the privilege of watching my son during the setup follow in my footsteps by trying to see how many bubbles he could light on his hand without burning himself.

"The hard part was finding minor Japanese celebrities willing to have their hands set on fire."

GIMME FIRE Lacking any minor Japanese celebrities available in Central Illinois, Gray rounded up these wacky and willing volunteers.

How I Did It

1. I filled a plastic sleeve with a mixture of hydrogen and butane, starting with butane from a can meant for filling cigarette lighters.

The idea to use a mixture of hydrogen and butane, rather than pure hydrogen, came from Yonemura Denjiro, the Japanese science educator I was paired with on the show. A neat trick for creating an accurate mixture, which I also learned from Denjiro, is to use a long, thin tube made of plastic sheeting (I used plastic sleeve material meant for packing posters, which I happened to have a lot of because I sell periodic table posters). Starting with a piece about six feet long, I put a mark about 1/4 of the way from one end. Then I filled the tube up to that mark with butane from a cigarette lighter refill can, keeping the rest of the tube pressed flat (like you flatten a toothpaste tube to bunch up all the toothpaste at the end). This measured out the right amount of butane, and then it was a simple matter to fill the rest of it with hydrogen to create the final mixture.

To make nice fine soap bubbles with the gas mixture, I used an aquarium aerator stone connected to the filled plastic sleeve with a hose and submerged in a tub of soapy water (a small amount of glycerin added to dish soap makes good bubbles). After squeezing out the gas like toothpaste from a tube, I had to work fairly fast to get the bubbles onto people's hands before they popped.

The key to not burning anyone with this demonstration is to make sure all the bubbles are on top of the hands, so the heat and flame will rise up away from the skin. If there are any bubbles underneath any hands, the flame will rise into the hand, causing singed hairs. Of course this always happens anyway, but fortunately the flame doesn't last long enough to do any real damage.

☠ REAL DANGER ALERT

Do not try this demonstration your-self. It's fire, and it's on your hand, and it is tremendously easy to burn yourself or set something on fire.

2. After adding the butane, I filled the sleeve up the rest of the way with hydrogen from a high-pressure cylinder.

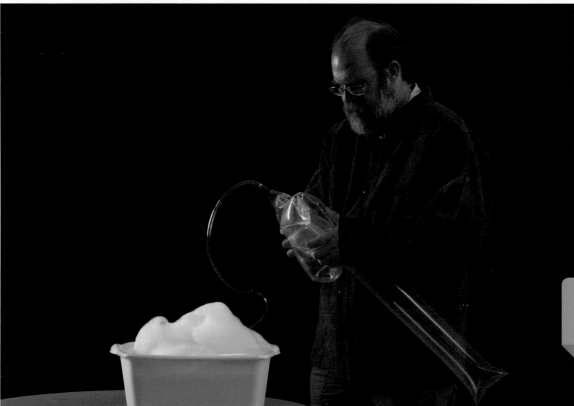

3. I created bubbles by squeezing the plastic sleeve out through an aquarium bubbler stone.

Fire Bird

Deep-frying a turkey can be a delicious Thanksgiving treat—or a deadly conflagration

OIL AND WATER DON'T mix: It's an old saying, but it's never more true than when you're talking about a pot of hot cooking oil and the moisture condensed on the surface of a frozen turkey. It's pretty incredible the amount of fire that simple combination can create.

Cooking oil is flammable, but it doesn't catch fire in a deep fryer because it never approaches the approximately 800°F required. Even if you drop a match in the fryer, the heat is conducted away from the flame and dissipates into the oil, and the fire goes out.

But oil dispersed into fine droplets is another beast entirely. Individual droplets heat up very quickly, and the burning of one drop creates enough heat to ignite the one next to it, and so on, making a cloud of oil droplets extremely flammable. Where might a cloud of oil droplets come from? From that big, frozen bird.

The recommended oil temperature for a deep fryer is 350°, well above the boiling point of water. When you drop food in, you immediately see bubbles; that is the water in the food boiling off. Put too much moisture in by lowering in a frozen turkey, and the vaporization of the water throws oil droplets into the air. A few of the droplets hit the burner under the pot and catch on fire, beginning a chain reaction that ignites a large cloud of droplets. The result is the smell of a county fair—and a towering inferno that can ignite everything around it.

So even if it's tempting to buy one of the many cheap turkey deep-fryers this time of year, you can add death by incineration to the other main reason not to: death by clogged arteries.

> "A few of the oil droplets hit the burner under the pot and catch on fire, beginning a chain reaction…"

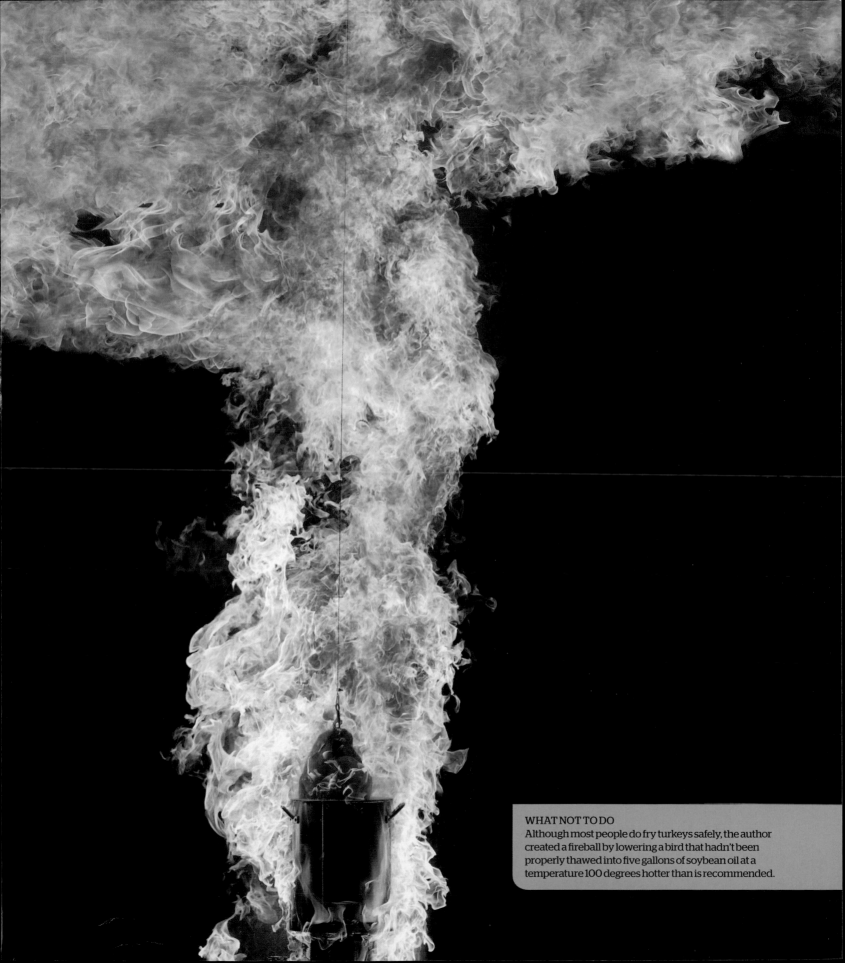

WHAT NOT TO DO
Although most people do fry turkeys safely, the author created a fireball by lowering a bird that hadn't been properly thawed into five gallons of soybean oil at a temperature 100 degrees hotter than is recommended.

How I Did It

As you can see from the picture, this demonstration creates a lot of fire. Literally thirty feet high, and very hot. To do it safely I needed to be able to raise and lower a 15-pound turkey from about 25 feet away. And to make the picture pretty, I didn't want to have any sort of bars or rods visible in the shot. I used two 10-foot pieces of pipe joined together and held up with steel cables. Then I looped a thin steel aircraft cable over a pulley fastened to the end of the pipes. At one end was a hook for hanging the turkey from, and at the other end, far away, was a handle I could pull on to raise and lower the turkey.

To simulate a very irresponsible turkey frying cook, I made two crucial

COOKING TIME Time-lapse photography shows how a plume of fine oil particles turns into a massive fireball in a second or two.

mistakes. First, I heated the oil way beyond its recommended frying temperature: Nearly 500°F. Second, I didn't thaw the turkey and intentionally let a lot of water condense on it from the surrounding air. This combination resulted in a tremendous amount of vigorous boiling, and resulting fine spray of oil.

People I show the pictures to assume the turkey must have been burned to a crisp, but actually it was nearly untouched: The fireball was huge, but only lasted a few seconds, and since the turkey went in frozen, it was didn't even begin to cook. We used three turkeys in getting these photos, which I gave to my assistant to finish cooking for her family. She used a regular oven.

Chemical Power

Out of Thin Air

A little oxygen is all a zinc-air battery needs to become a powerhouse

A BATTERY THAT RUNS on air? Why, that's almost as good as a car that runs on water! Those cars are fantasy, but batteries that run on air are actually quite common, especially among older people. Tiny zinc-air batteries are widely used in hearing aids, where they have replaced toxic mercury-based batteries in providing a small but steady stream of power. They supply more energy for their size than any other battery, because they draw some of their power straight from the air.

All batteries generate power with two chemical reactions: one that produces electrons at the anode (negative terminal) and one that absorbs them at the cathode (positive terminal). This creates a circulation of electrons—an electrical current—from the anode to the cathode. Most batteries contain all the chemicals needed for both reactions.

But zinc-air batteries contain only the anode-side chemical, zinc metal, which is converted to zincate ions and then zinc oxide. This releases two electrons per atom of zinc, which are absorbed by oxygen on the cathode side.

Zinc-air batteries can pack more power into a smaller space than other batteries for the same reason that jets run for longer than rockets. Rockets, which must operate in the vacuum of space, have to carry both fuel and oxygen to burn it with. Jets need to carry only the fuel, since they can pull in oxygen from the air.

The downside is that jets can't produce as much thrust as rockets, because there's a limit to how fast they can suck in air. Zinc-air batteries have the same limitation. They can deliver a large amount of energy, but only relatively slowly; they're like endurance runners, not sprinters—the tortoise to Energizer's bunny.

How I Did It

There isn't really much technique to doing this demonstration, it's just showing a battery running a toy. Taking apart a zinc-air watch battery, on the other hand, is a real pain in the rear. I had to disassemble several of them before I had each individual component undamaged.

A. Cathode can with air holes
B. Teflon air filter
C. Graphite-and-wire-mesh air cathode
D. Dielectric filter paper
E. Zinc-powder anode
F. Anode can
G. Insulating ring

WHAT'S NEXT Because zinc-air batteries must be open to the air, the water inside them eventually evaporates, limiting their life span. But future cells could use electrolytes made of ionic fluids, which do not evaporate. The potential exists for rechargeable zinc-air batteries with 10 times the capacity of today's cells.

"Zinc air batteries are like endurance runners, not sprinters."

ELECTRIC FLYER These zinc-air button cells aren't powerful enough for liftoff, but they can keep the rotors turning longer than other batteries can.

Apple Juice

Charge your gadgets with a piece of fruit and some pocket change

ARTHUR C. CLARKE WROTE that "any sufficiently advanced technology is indistinguishable from magic," but he was wrong. It's easy to tell the difference—technology works. For example, "remote-viewing" mentalists claim they can see events far away, yet they fail every test. In fact, remote viewing is simple: It's called TV.

Another example that recently circulated online was a fake video of someone charging his iPhone by jamming the end of a USB cable into an onion. How do I know it was fake? First, you need contacts made of two different metals, and second, you can't get enough voltage out of a single vegetable. What makes the ruse so disappointing is that it is possible to charge an iPhone this way, if you do it right.

A regulation vegetable battery, made by sticking strips of zinc and copper into a potato, generates about half a volt. The electricity comes from the oxidation of zinc; the vegetable is just an electrolyte (conductive barrier), and the copper completes the circuit. Stacking alternating layers of vegetable, zinc and copper is like wiring batteries in series, each set adding its voltage to the total.

After some 10 volts' worth of teary-eyed onion peeling, I decided to switch to apples, using a fruit corer to cut out the apple rods and a cheese slicer to cut them into disks. Pennies with the copper plating sanded off on one side made a handy source of copper and zinc layers in one.

About 150 of these, arranged into six parallel batteries of 25 apple/zinc/copper layers each, yielded enough power to charge an iPhone, but only for about a second. (Much larger zinc plates and whole slices of apple would have provided more power for longer.) Around 200 of the layers went into one three-foot-long apple battery, delivering much higher voltage. I was able to create a visible, and potentially fatal, spark with this battery. Yes, in the right configuration, you can electrocute yourself with an apple.

CORE WORK The first step in creating a high-voltage fruit battery: Coring an apple.

QUICK CHARGE Six apple-penny batteries wired in parallel, each with 20 to 25 cells (apple slices) is just barely enough to charge an iPhone for about a second.

"I was able to create a visible, and potentially fatal, spark with this battery."

How I Did It

This was one of the most involved demonstrations I've done, due to the need to produce a couple hundred half-sanded pennies. Post-1982 pennies are made of copper-plated zinc, so if you sand off the copper plating on one side of them, and combine them with solid copper pre-1982 pennies, you have both metals needed to create a battery.

I mounted a short iron water pipe, whose inside diameter was just about the same as a penny, up against a small vertical belt sander, leaving a gap between the end of the pipe and the sanding belt that was just a bit thinner than a penny. Then I filled the pipe with a stack of pennies, and pressed them up against the running belt with a plunger. Each penny in turn was sanded down until it was thin enough to fit through the gap, at which point the moving belt threw the penny out of the machine, allowing the next one to advance into position. It worked surprisingly well as an automatic penny sander, plopping out a sanded penny every few seconds.

I first thought of using onions because I figured I could core them with a fruit coring tool (basically a very thin-walled metal tube with a sharpened edge, which you can use to cut plugs out of fruits), and then the layers would separate into lots of individual disks. Unfortunately onion didn't work very well, I think because each layer has a sort of membrane on one side that doesn't conduct electricity very well. So I switched to using apples instead, and had to manually slice the cores into disks.

FRUIT POWER Coring and slicing yields penny-size apple disks. About 200 in series with pennies produces enough voltage to charge a capacitor and create a spark.

 ## REAL DANGER ALERT

This experiment could damage your iPhone if done improperly.

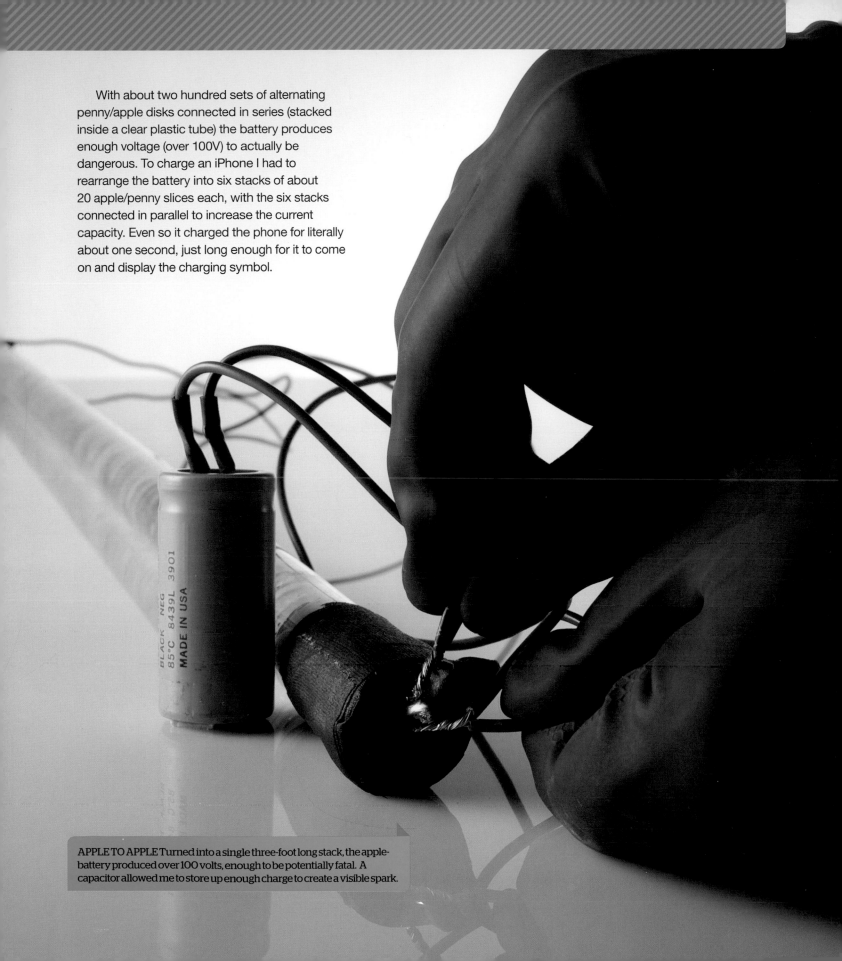

With about two hundred sets of alternating penny/apple disks connected in series (stacked inside a clear plastic tube) the battery produces enough voltage (over 100V) to actually be dangerous. To charge an iPhone I had to rearrange the battery into six stacks of about 20 apple/penny slices each, with the six stacks connected in parallel to increase the current capacity. Even so it charged the phone for literally about one second, just long enough for it to come on and display the charging symbol.

BLACK NEG 3901
85°C 8439L
MADE IN USA

APPLE TO APPLE Turned into a single three-foot long stack, the apple-battery produced over 100 volts, enough to be potentially fatal. A capacitor allowed me to store up enough charge to create a visible spark.

The Other White Heat

You know bacon is delicious, but did you know it has enough energy to melt metal?

I RECENTLY COMMITTED MYSELF to the goal, before the weekend was out, of creating a device entirely from bacon and using it to cut a steel pan in half. My initial attempts were failures, but I knew success was within reach when I was able to ignite and melt the pan using seven beef sticks and a cucumber.

No, seriously. The device I built was a form of thermal lance. A thermal lance, typically made of iron instead of bacon, is used to cut up scrap metal and rescue people from collapsed buildings. It works by blowing pure oxygen gas through a pipe packed with iron and magnesium rods. These metals are surprisingly flammable in pure oxygen, releasing a huge amount of heat as they are consumed. The result is a jet of superheated iron plasma coming out of the end of the pipe. For sheer destructive force, few tools match a thermal lance. But iron isn't the only thing that's flammable in a stream of pure oxygen.

Bacon is fattening because it contains a lot of chemical energy tied up in its proteins, and especially in its fat. You can release that energy either by digesting it or by burning it with a healthy supply of oxygen. The challenge isn't creating the heat; it's engineering a bacon structure strong enough to withstand the stress of a 5,000°F bacon plasma flame.

I used prosciutto (Italian for "expensive bacon") because it is a superior engineering grade of meat. I wrapped slices of it into thin tubes and baked them overnight in a warm oven to drive off all the water. Then I bundled seven of those together, wrapped them in additional slices, and baked the bundle again until it was hard and dry.

To make an airtight, less-flammable outer

> **"Bacon contains a lot of chemical energy tied up in its proteins, and especially in its fat."**

GREASE FIRE Pure oxygen flows from a metal pipe through a core of baked prosciutto, generating a grease fire hot enough to ignite steel and burn a hole clear through this pan. A wrapping of less-flammable uncooked prosciutto focuses the flame into an intense bacon-plasma torch.

casing, I wrapped this fuel core with uncooked prosciutto before attaching one end of it to an oxygen hose. You can't imagine the feeling of triumph when I first saw the telltale signs of burning iron: sparks bursting from the metal, and then a rush of flame out of the other side as I witnessed perhaps the first-ever example of bacon-cut steel. And the lance kept on burning for about a minute.

It turns out there are much easier ways to do this. For example, while researching how to build a vegetarian lance, I hit on the perfect pipe material—hollowed-out cucumbers. The pressure-containment capacity of a standard cucumber is remarkable, and the smooth skin makes it easy to create an airtight seal with the pipe delivering oxygen to the device. A cucumber packed with beef sticks will burn for almost two minutes, and a completely vegetarian version stuffed with breadsticks, though not quite as long-lasting, still produces a very impressive flame.

The lesson here is that food is a source of serious amounts of energy. Pure oxygen helps release it in a much shorter time than usual, but it's really the chemical energy in the bacon that makes the steel pan burn. Whether it's worth building a bacon lance to demonstrate this—well, only you can be the judge of that.

How I Did It

This is definitely both the most fun and the most pointless of all the demonstrations I've ever done. It absolutely does work: You can cut steel with bacon (technically prosciutto) and oxygen gas. I've done it four or five times now, and it never fails.

The key is creating a structure for the oxygen gas to flow through which is both air tight, and which exposes a large surface area of prosciutto to the gas as it's flowing by. I decided to do this by making a bundle of thin tubes.

Prosciutto is flexible when raw, but turns stiff and hard if you bake it. I greased up some 1/4" diameter steel rods and wrapped individual sheets of prosciutto around them. Then I baked the rods for 2-3 hours at around 250°F, resulting in stiff prosciutto tubes after I pulled them off the rods. I bundled seven of these together (one in the middle and six around it), and held them together by wrapping them with more prosciutto. After baking again, the whole thing was rigid and felt quite solid.

But baked prosciutto is not air tight: Cracks develop all over. So I wrapped the whole thing with several layers of fresh prosciutto, the goal of course being to use nothing but bacon/prosciutto to build the device.

With the device ready, the next step was to connect it to a source of oxygen, which I did using a rubber pipe sleeve and duct tape. Pipe sleeves are commonly available in hardware stores, and consist of a short rubber tube with hose clamps on either end, designed to slip over a pipe and provide a flexible link or leak repair.

Knowing the power of pure oxygen, I was fairly nervous the first time I lit a bacon lance, but it seemed to be pretty stable and controllable. It burned for a minute or less before flame burst out the side, or reached the rubber joint, which then of course also starts burning. But in that time they easily provide enough heat to ignite and cut steel.

MEAT STICKS I wrapped slices of prosciutto around fiberglass rods, baked them dry, and bundled seven tubes into a bacon fuel core.

REAL DANGER ALERT

Theodore Gray is trained in lab safety and flammable meat handling. Don't try this at home.

BACON POWER You can see multiple jets of pork-plasma emerging from this fully operational bacon lance.

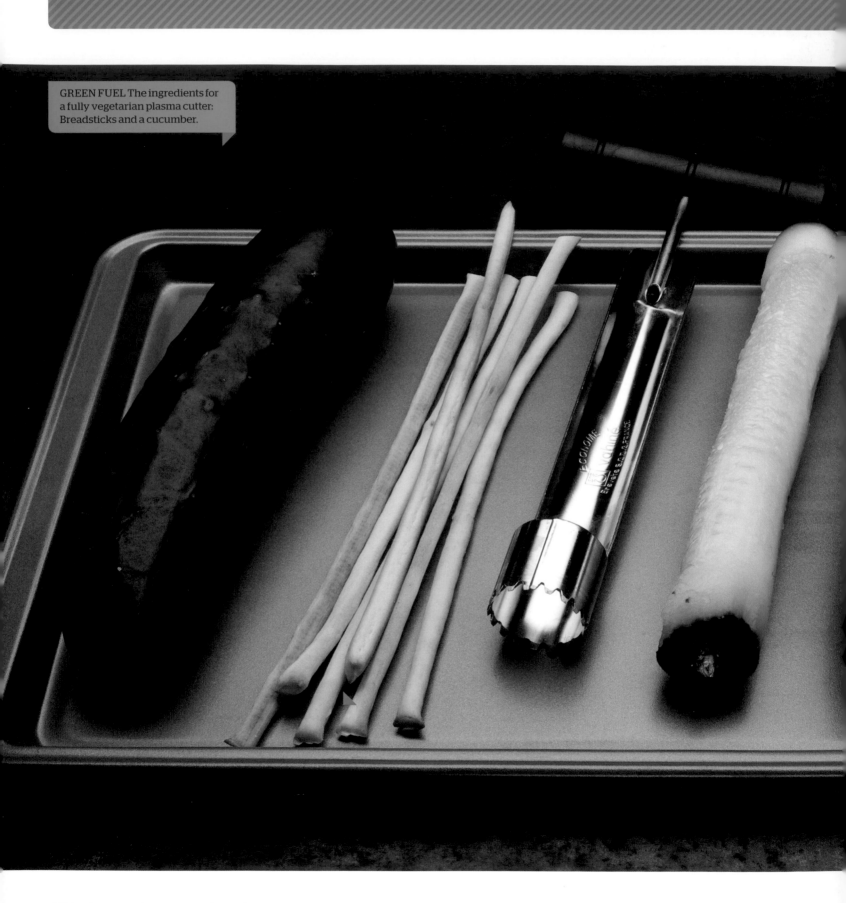

GREEN FUEL The ingredients for a fully vegetarian plasma cutter: Breadsticks and a cucumber.

GREEN LANTERN A cucumber makes an even better edible thermal-lance housing, since its outer rind contains the pressure of the very hot flame without burning up.

Stopping Power

How the same trick recovers energy in a hybrid car and can help you survive the zombie apocalypse

WHEN YOU STEP ON your car's gas pedal, you're taking chemical energy stored in gasoline molecules and turning it into the kinetic energy of a moving car. It's not the most efficient process in the world, but much of the energy content of the gas ends up as useful motion. Yet when you step on the brakes, you're just throwing it all away—the kinetic energy is wasted as heat in the brake pads and rotors.

Electric cars can recover that energy by taking advantage of a wonderful property of electric motors: They also work as generators.

When you're accelerating in an electric car, the battery forces electrons through wire coils in the motor. That generates magnetic fields, which push against permanent magnets in the motor, creating a force on the motor shaft that turns the wheels.

This chain is reversed when you brake. Forcing a magnetic field through a wire produces a flow of electrons. When you cut the power to a spinning motor, the coils are still moving through the magnetic field of the permanent magnets, so the coils immediately start generating a current, which you can use to charge the battery. Drawing energy out of the motor causes a drag on the shaft, which creates a braking force on the wheels.

The perfect device to demonstrate this is a 12-volt car winch with a manual crank handle. After removing the tow cable, you can spin the motor by cranking the handle, easily building up enough power to operate a 120-volt AC power inverter. Hooked up to an exercise bicycle, a winch would make a great emergency generator in case of a zombie apocalypse.

> "Hooked up to an exercise bicycle, a winch would make a great emergency generator in case of a zombie apocalypse."

HANDY POWER
A car winch cranked
in reverse generates
enough electricity
to run this plasma globe.

How I Did It

I wanted to do a demonstration of regenerative braking by showing that an electric motor can work in reverse as a generator. The question was, what kind of motor could I get that I could spin at high speed with a hand crank? It occurred to me that a car winch should have the right properties: A strong motor geared down to turn a spool slowly. Gear chains also work both ways, so if a fast-spinning motor turns the spool slowly, then turning the spool slowly will spin the motor fast.

It turned out that the gearing ratio on the winch I got was too high: I was not strong enough to turn the spool. But, happily it was easy to remove one of the links in the gear chain, reducing the gearing ratio to the point where it was just about right: Vigorous cranking of the handle spun the motor at the right speed to generate the 12V needed to run a power inverter to create 120V.

If I'm ever caught in a zombie apocalypse and need power to run my iPhone, I think this is the most practical method. Just about any old electric motor will do, and if it were connected to something like an exercise bicycle so you could turn it with your feet, you could generate very usable amounts of energy. Generating 100 Watts with your legs for an hour on end is not unrealistic if you're in reasonable shape. (But to give you an idea of how incredibly spoiled we are, buying from the electric company the same amount of energy as you produce in an hour of vigorous exercise would set you back about one cent.)

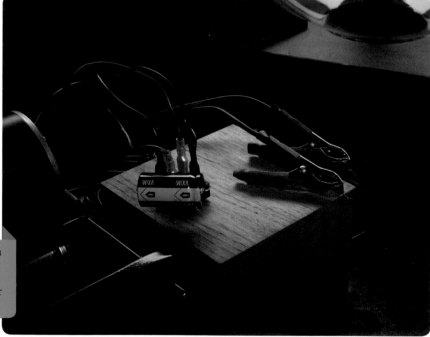

POWER TO THE PROJECT A full wave bridge rectifier and a capacitor turn the AC power generated by the winch motor into roughly 12V of rough DC power suitable for running this plasma globe (which normally operates off a 12V power supply).

CRANKING OUT ART
This plasma sculpture
runs off 120V AC, so Gray
connected a car power
inverter to the winch:
It converts 12V DC into
120V AC.

Index

A

acetone, 60, 61–65
acetylene gas, 18, 19, 20
air
  canned, 44, 45–46
  fire with compressed, 126, 127–129
  graphite-and-wire-mesh air cathode, 164
  methane gas and, 56, 124
  Teflon air filter, 164
  zinc-air batteries, 164–165
alkali metals, 150, 151, 152–153
Alla-Seltzer, 40
alloys, 24
aluminum cages, for magnets, 27
ammonium nitrate/fuel oil (ANFO), 126
ammunition, 134–135, 136
ANFO. *See* ammonium nitrate/fuel oil
anode, 164
apple battery, 166–169
Aspartame, 40, 42

B

bacon thermal lance, 170, 171–173
batteries
  apple, 166–169
  electrons of, 164
  fruit, 166–169
  making fruit, 168–169
  pennies in, 166–169
  vegetable, 167
  zinc-air, 164–165

Becquerel, Henri, 75
bicycle
  as generator, 176, 178
  locks, 44, 45–47
bismuth, 80, 81
boron, 28
borosilicate glass, 28
Brainiac, 151
braking, 176
bubbles, on fire, 154, 155–157
bullets, 104, 134–135, 136
butane gas, hydrogen and, 154, 155–157

C

C4, 134
CaC2. *See* calcium carbide
calcium carbide (CaC2)
  cannons, 20
  on ice cubes, 21
  miner's lamp, 18, 19
camera
  flashbulbs, 134, 136, 137
  X-ray, 74, 75, 76–79
canned air, 44, 45–46
cannons, calcium carbide, 20
capacitor, 178
carbide miner's lamps, 18, 19
carbon, 30
cartridge, ammunition, 134–135, 136
catalysts, 60, 61–65
cathode, 164
cesium, 151
charcoal, 146, 148
chemical power, 176
citric acid, 40

Clarke, Arthur C., 166
cocaine, 28
coffee, self-heating, 42
combustion, 138, 139–141
compressed air, fire with, 126, 127–129
conspiracy theories, 138
cooking oil, on fire, 142, 158, 159–160
cooling
  bike locks, 44, 45–47
  liquid nitrogen, 84, 85–87
  mercury, 102, 108, 109–113
  rubber balls, 48–49
copper, 60, 61–65
cotton, 138
cucumber thermal lance, 172, 174–175

D

Davy, Humphry, 124
Denijiro, Yonemura, 156
detonation, 126, 134
diamonds, 30, 31, 32, 33–35, 44
dielectric filter paper, 164
diodes, light-emitting, 16
diphenyl oxalate, 66, 67–73
Drano, 66, 67, 69–73
dry ice, 59
dye, phosphorescent, 66, 67–73

E

ECM. *See* electrochemical machining
electric cars, 176

Index

Index